Martin Horauer

Clock Synchronization in Distributed Systems

Martin Horauer

Clock Synchronization in Distributed Systems

Architecture and Evaluation of Ethernet-based Network Interfaces with support for precision clock synchronization

Südwestdeutscher Verlag für Hochschulschriften

Impressum/Imprint (nur für Deutschland/ only for Germany)
Bibliografische Information der Deutschen Nationalbibliothek: Die Deutsche Nationalbibliothek verzeichnet diese Publikation in der Deutschen Nationalbibliografie; detaillierte bibliografische Daten sind im Internet über http://dnb.d-nb.de abrufbar.
Alle in diesem Buch genannten Marken und Produktnamen unterliegen warenzeichen-, marken- oder patentrechtlichem Schutz bzw. sind Warenzeichen oder eingetragene Warenzeichen der jeweiligen Inhaber. Die Wiedergabe von Marken, Produktnamen, Gebrauchsnamen, Handelsnamen, Warenbezeichnungen u.s.w. in diesem Werk berechtigt auch ohne besondere Kennzeichnung nicht zu der Annahme, dass solche Namen im Sinne der Warenzeichen- und Markenschutzgesetzgebung als frei zu betrachten wären und daher von jedermann benutzt werden dürften.

Verlag: Südwestdeutscher Verlag für Hochschulschriften Aktiengesellschaft & Co. KG
Dudweiler Landstr. 99, 66123 Saarbrücken, Deutschland
Telefon +49 681 37 20 271-1, Telefax +49 681 37 20 271-0, Email: info@svh-verlag.de
Zugl.: Wien, TU, Diss., 2004

Herstellung in Deutschland:
Schaltungsdienst Lange o.H.G., Berlin
Books on Demand GmbH, Norderstedt
Reha GmbH, Saarbrücken
Amazon Distribution GmbH, Leipzig
ISBN: 978-3-8381-0334-1

Imprint (only for USA, GB)
Bibliographic information published by the Deutsche Nationalbibliothek: The Deutsche Nationalbibliothek lists this publication in the Deutsche Nationalbibliografie; detailed bibliographic data are available in the Internet at http://dnb.d-nb.de.
Any brand names and product names mentioned in this book are subject to trademark, brand or patent protection and are trademarks or registered trademarks of their respective holders. The use of brand names, product names, common names, trade names, product descriptions etc. even without a particular marking in this works is in no way to be construed to mean that such names may be regarded as unrestricted in respect of trademark and brand protection legislation and could thus be used by anyone.

Publisher:
Südwestdeutscher Verlag für Hochschulschriften Aktiengesellschaft & Co. KG
Dudweiler Landstr. 99, 66123 Saarbrücken, Germany
Phone +49 681 37 20 271-1, Fax +49 681 37 20 271-0, Email: info@svh-verlag.de

Copyright © 2009 by the author and Südwestdeutscher Verlag für Hochschulschriften Aktiengesellschaft & Co. KG and licensors
All rights reserved. Saarbrücken 2009

Printed in the U.S.A.
Printed in the U.K. by (see last page)
ISBN: 978-3-8381-0334-1

Abstract

A system-wide global time base with known precision is of pivotal importance for the design and operation of distributed systems as well as an enabling technology for applications like location-based services. The increasing requirements of these driving applications and the large scale of the underlying systems demand clock synchronization down to the *ns*-range. To date, for many applications this cannot be established with present software synchronization strategies; specialized hardware support and the use of GPS-timing receivers is mandatory. The applicability of these solutions, however, is limited by the high cost for the additional, dedicated cabling and the antennas for the GPS receivers, which require clear-view of sky for proper operation.

Recently the IEEE approved the 1588 standard for a precision clock synchronization protocol for networked measurement and control systems. By equipping existing computer networks with moderate hardware extensions at the network interfaces and a standard protocol software stack, an average precision below the μs-range can be achieved. Independently from the balloting process and based on relevant scientific literature the research project SynUTC established a clock synchronization framework with sound theoretical concepts and well engineered hard- and software.

This thesis proposes an architecture for network interfaces and networked devices that will render a worst-case precision in the *100 ns*-range possible. The proposed mechanism, which is applicable for any packet-oriented data network, inserts time information into data packets at the interface between the physical layer transceiver and the network controller upon packet transmission and reception, respectively. Local time is supplied by a high-resolution rate-adjustable adder-based clock, which also contains hardware support easing interval-based external clock synchronization, like maintaining time and accuracy intervals and interfaces to GPS receivers. This architecture allows an improvement of at least an order of magnitude over other existing solutions; it is accomplished by small modifications of existing commercial-off-the-shelf devices, without impairment of their original functionality. Part of the principle of operation is verified with a prototype implementation that was also used in conjunction with other devices for an experimental evaluation. The results of the presented experiments validate the proposed techniques and reveal actual values for the worst-case precision that might be achieved. The presented solution provides a synchronization that can otherwise be achieved only with the help of specialized GPS timing receivers, thus excellently complementing these solutions when increased fault-tolerance is required or when access to an antenna is not feasible.

Contents

Abstract   i

1 Introduction   **1**
    1.1 Clock Synchronization Strategies 1
    1.2 Application Domain 3
    1.3 Outline 8

2 State of the Art of Clock Synchronization   **11**
    2.1 System Modelling 12
       2.1.1 Clocks and Processors 13
       2.1.2 Communication Subsystem 14
       2.1.3 Faults 15
    2.2 A taxonomy of clock synchronization algorithms 16
       2.2.1 Structure of clock synchronization algorithms 16
       2.2.2 Clock synchronization building blocks 17
    2.3 Requirement analysis 27
       2.3.1 Clock Properties 27
       2.3.2 Clock Reading Error 33
       2.3.3 Clock Granularity an Clock Rate Adjustment 35
       2.3.4 Coupling to an External Reference Time 35
    2.4 Summary 36

3 Related Work   **39**
    3.1 MARS - The Maintainable Real-Time System 40
    3.2 The Time-Triggered Protocol 42
    3.3 The Network Time Interface 45
    3.4 IEEE Standard 1588 47
    3.5 Summary 49

4 Network interface architectures supporting tight clock synchronization   **51**
    4.1 System Architecture 53
    4.2 Network interface for End-systems 57
       4.2.1 Clock synchronization support for Network Interface Cards . . . 58
       4.2.2 Prototype: MII-NTI 61
    4.3 Clock architecture 63

		4.3.1	Bus Interface and Timestamp Unit	64
		4.3.2	Local Time Unit	65
	4.4	Networked devices		68
		4.4.1	Clock synchronization support for Switches	68
		4.4.2	Switch Add-On	72
	4.5	Summary		73

5 Delay variations of the Physical Layer — 75

	5.1	Models of the physical communication link		75
		5.1.1	Cable model	75
		5.1.2	10 Base-T Physical Layer Devices	77
		5.1.3	100 Base-Tx Physical Layer Devices	78
	5.2	Evaluation		82
		5.2.1	Evaluation System Hardware	83
		5.2.2	Evaluation System Software	83
		5.2.3	Evaluation System Setup	84
	5.3	Measurement Results		86
		5.3.1	Direct connection	86
		5.3.2	Networked devices	98
	5.4	Summary		103

6 Conclusion and Future Work — 107

Bibliography — 109

Appendix — 118

Glossary and Abbreviations — 120

Chapter 1
Introduction

> Leslie Lamport's definition of a distributed system: *"You know you have one when the crash of a computer you've never heard of stops you from getting any work done."* [72]

A distributed system is a collection of autonomous computers linked by a computer network and supported by software that enables the collection to operate as an integrated facility. It is very easy to understand why these systems are popular. They allow the sharing of information and resources over a wide geographic spread and they are usually better than traditional centralized systems in terms of sharing, cost, growth and autonomy. In contrast, the above citation from Leslie Lamport states clearly that there are still some short-comings and weaknesses with existing implementations. The distributed nature of these systems has to cope with unreliable and insecure communications and independent failures. These problems become aggravated when the system is operating critical real-time applications such as aerospace systems, life support systems, nuclear power plants, drive-by-wire systems and computer-integrated manufacturing systems. Common to all these applications is the demand for maximum reliability and high performance from computer controllers, since a single controller failure in these applications can lead to disaster. Next to these an increasing number of distributed applications, such as process-control applications, transaction processing applications, or communication protocols, rely on autonomous computers that need to cooperate for initiation of actions or recording of events. Therefore, causal ordering is often required, a means that can be provided with the help of synchronized clocks so that every computer node has approximately the same view of time. In addition, when synchronized clocks are at hand the performance of a distributed system can be improved by reducing communication, see [59] for some practical uses. Other practical applications and uses will eventually emerge when tight clock synchronization becomes available at affordable costs. Therefore, the aim of this thesis is to analyze and underpin concepts for the improvement of existing hybrid clock synchronization mechanisms within a distributed system.

1.1 Clock Synchronization Strategies

Clock synchronization may be achieved either by hardware or by software. A coarse classification of typical clock synchronization mechanisms is given in Fig. 1.1 depending on the required precision and the geographic spread of the distributed system.

- The Network Time Protocol (NTP) is a network protocol and a collection of algorithms for synchronizing computer clocks over packet networks and is mainly used throughout the Internet. Computers synchronize their local time to those on remote time-servers that are assumed to have the correct time, for details see [68] and [69]. Under some realistic conditions, maximum errors of approximately *20 ms* were observed, cf. [105].

- For local area networks with soft real-time requirements synchronization in the range of some few *ms* is acceptable. Software clock synchronization algorithms use standard communication networks and send synchronization messages to get the clocks synchronized. Numerous solutions have been presented, analyzed and evaluated; see [82] or [100] for an overview and [114] for a comprehensive bibliography.

- When computer nodes are equipped with moderate hardware support, the software clock synchronization algorithms can yield synchronization tightness in the range of several μs. Such a mechanism was implemented in the MARS project, see [53], and a similar hybrid scheme of software synchronization with hardware assistance for homogeneous distributed systems — although for a different network architecture — can be found in [81].

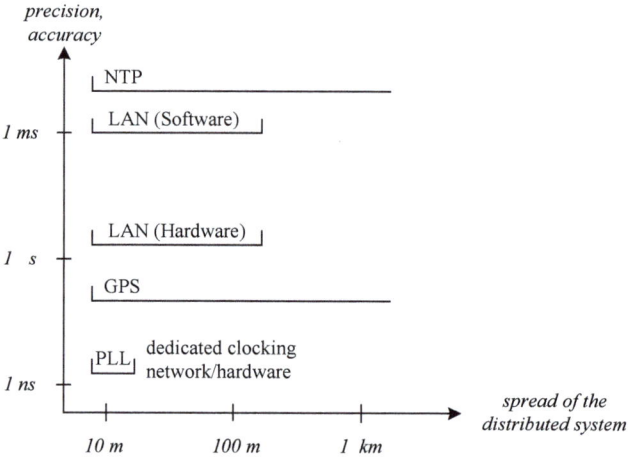

Figure 1.1: Clock Synchronization classification

Our research project *Synchronized Universal Time Coordinated* (SynUTC) pursued the problem of how to establish an accurate common notion of time among the nodes of a distributed system relying on the same hybrid mechanism. The local clocks of every node are kept within a few μs of each other by solely exchanging dedicated network packets. By incorporating an external time source like a *Global Positioning System* (GPS) receiver the local time at every node is kept within a few μs of *Universal Time Coordinated* (UTC), the only official and legal standard time, as well. The recently established IEEE 1588 Standard that proposes a similar mechanism for tight clock synchronization also fits into this category, see [40].

- When every node can be equipped with a dedicated GPS Receiver clock synchronization in the range down to *100 ns* is possible, although such a pure coupling of common-of-the-shelf (COTS) receivers without additional means for clock validation is inadequate for safety critical applications as these devices exhibit some rare failures, see [29].

- Dedicated hardware support can yield synchronization accuracy down to several *ns*. These solutions require in general a dedicated clocking network along with appropriate synchronization hardware which operates similar to the phase-locked-loop technique. The additional cabling and hardware is only affordable for distributed nodes that are located within a few meters, as is the case, e.g., in multiprocessor systems [46].

The topics of this thesis address the problem of how to enhance the hybrid LAN-based clock synchronization architectures with moderate hardware support in order to enhance the achievable synchronization accuracy, in particular how to keep the nodes clocks within a distributed system bounded within some *ns* towards each other (= *precision*) and towards an external reference time (= *accuracy*), respectively. The presented work is very closely related to the newly established IEEE 1588 Standard and illustrates some deficiencies of this standard and how these could be overcome.

1.2 Application Domain

Since clock synchronization is used throughout the spectrum of distributed systems starting from a single VLSI chip, and ranging up to a global network it is conceivable that the effect of even a slight improvement in the tightness of synchronization may be sweeping. For example, tighter synchronization of the transmitting and receiving endpoints of communication links can lead to better utilization and hence larger throughput of the communication network; better synchronization may imply shorter processing time for large databases. In order to state the benefits and usefulness of tighter clock synchronization in a distributed system more precisely, we present some examples that may benefit from such an enhanced solution. In particular, we give some examples that could be applied in such different domains as mobile communications or power systems in order to show the range of applicability.

Location Services: With the advent of mobile communications a huge marketplace has been created where stiff competition has caused mobile technology to improve rapidly. New services are constantly being introduced (e.g., short message service in *Global System for Mobile* (GSM)) to capture larger market shares.

In recent years it has become apparent that there is a large demand for mobile location services. The service could provide a range of functions, such as car navigation, fleet management, location charging (e.g., road pricing) or advertising and anti-theft devices. These services currently available use their own communication systems and radio frequency allocation — a pooled resource would be far more efficient and cost effective. In 1997 the U.S. Federal Communications Commission introduced a mandate to enforce all mobile telephone networks to provide mobile locations for emergency services [13]. This mandate requires that cellular, personal communication services and specialized mobile radio service providers deploy a means of automatically locating emergency callers to within 125 m in 67 % of all measurements by October 31, 2001.

Unfortunately, there are a number of challenges to overcome to implement such a service. The accuracy of such a service is probably not as great as GPS, the satellite based location system — *differential-GPS* is accurate to some few meters. However, the mobile solution would be inexpensive by comparison. It does not require a direct line of sight communication and can penetrate buildings. It would, therefore, be aimed at a different market. The third generation mobile standard, *Universal Mobile Telecommunications System* (UMTS), has recently been developed by the European Telecommunications Standard Institute. There is currently much research and development being carried out with the collaboration of major mobile communications companies. In this way UMTS networks are designed with location estimation services in mind.

There are several methods that can be used to calculate an unknown mobile position from measurements based on signals from base stations of known position, see [84], [115] and [83] for an overview. These proposed location technologies fall into two broad categories: Network-based solutions and handset-based solutions.

Network-based Location Services:

- The *Signal Strength Analysis* works by measuring the signal strength of the mobile station at at least three base stations. This measurement is then directly related to the separation distances between the mobile and the base stations. The conversion from signal strength to distances and fading problems need to be overcome here in order to provide the required accuracies.

- The *Angle of Arrival* (AOA) technique relates the absolute angle of arrival of the signal of the mobile station at two or three base stations. This technique relies on antenna arrays which provide the direction finding capability to the receiver. This method has some impracticalities due to the size, alignment and array separation problems of the antenna array. Field trials in London, see [76], revealed some problems due to the achievable accuracy.

- The *Time of Arrival* (TOA) Technique is enforced by bouncing a signal back between the mobile and the base station in either direction. With the knowledge of the propagation time and the measurement of three such data sets to different base stations one can easily triangulate the mobile position. The required duplex signal transmission is one major drawback of this approach.

- *Time Difference of Arrival* (TDOA) measures the relative arrival time of the signal from the mobile at three base stations. Precise clock synchronization of the base stations will be required for this method.

Some of these network-based solution can be applied in the reverse direction for handset-based methods as well, but due to size and reasons of practicality, only a modified scheme of TDOA is considered.

Handset-based Location Services:

- *GPS* chip-sets integrated into a mobile allow for direct location estimation in the range of 5 to 40 m accuracy since the US government removed the Selective

Availability mask in May 2000. The latter was used in the past to reduce the location estimation accuracy. In general, GPS now is the most popular radio navigation aide and has overtaken virtually all other forms of radio navigation because of its high accuracy, world wide availability, and low cost. Problems that need to be overcome with GPS handset-based solutions are the relative long time to first fix, and the requirement for a clear view of the sky.

- Network *Assisted GPS* uses fixed GPS receivers that are placed at regular intervals to fetch data that can complement the readings of the mobile GPS receiver. The assistance data makes it possible for the receiver to make timing measurements from the satellites without having to decode the actual messages and thus reduces greatly the time needed for a GPS receiver at the mobile to calculate the location. This solution reduces the time to first fix, the problem of the above solution but the requirement for a clear view of the sky is still pertinent, especially in urban locations.

- *Enhanced Observed Time Difference* (E-OTD) in GSM and *Idle Period Downlink* (IP-DL) in UMTS are TDOA variants for handset-based solutions. Here the mobile listens to bursts from multiple base stations and measures the time difference. These measurements are used to triangulate the actual mobile location. This requires that the base station positions are known and that the data sent from different sites is synchronized. The most common way of synchronizing the base stations is via the use of fixed GPS receivers. The calculation can then either be done in the mobile or at dedicated network nodes. The accuracy based on the achievable time synchronization is expected to be below 125 m, and unlike GPS these methods are not reliant on a clear view of the sky.

Hybrid techniques using more than one of the above techniques have been suggested in order to improve the location estimate accuracy. Currently TDOA variants are considered the leading candidates for any future location system. The precise synchronization of the base stations therefore will be provided with the help of GPS, since there seems to be no other solution at hand that can provide the required precision. In fact, the clock synchronization approach laid out in this thesis could be a promising alternative candidate to provide tight synchronization since the network infrastructure between the base stations is already present. A combination of both systems would very well complement each other and avoid spurious problems infrequently encountered with some GPS receivers, see [29].

Fault location in Power Grids: Another field of application is power distribution systems. Tight clock synchronization could be used to provide means for on-line fault detection and and estimation of the fault location [47]. Reliable provision of electrical power is of utmost importance for our daily human life. This in turn requires redundant, reliable power system components that can be easily maintained and exchanged, see [60]. Disregard of these criteria can have catastrophic consequences, as they were experienced after recent power outages in the US[1], in Europe[2] and in New-Zealand[3]. Due to recent deregulations within the energy sector and poor environmental acceptance an improvement of reliability with the help of building-up additional power lines or other power system components is limited. A practical approach to improving the quality of power supply provision is to reduce the time between fault occurrence and resumption of operation. In case of an interruption the following steps need to be undertaken:

1. fast and reliable fault detection

2. selective switch-off of defect power system components

3. exact fault location

4. repair of defect power system components and resumption of operation

Fault-detection and selective switch-off are currently ensured with the help of present distance protection techniques. For fault location on-line and off-line methods are used where only current off-line methods provide exact location accuracy, especially when underground cables are considered.

On-Line fault location techniques:

- Methods relying on measurement of impedances are currently built-in into existing distance protection relays and provide a coarse estimation of the fault location. This information satisfies the needs for a selective switch-off of defect power system components immediately after the occurrence of a fault. The achievable precision for pin-pointing the fault location is in the range of 3 to 10 %, see [54], or [7], and is therefore insufficient for an exact fault location, especially when underground power cables are affected.

- Some other methods are based on travelling waves theory. A fault location technique that uses the propagation delays of the first wave that is emanated after a surge is affecting the cable is presented in [42]. They measure the time of arrival of this wave at both ends of the cable and calculate the fault location using the a priori known cable length. Field tests in Tokio revealed a location error within 1 % of the cable length. This method requires additional measurement cabling from both ends fed together in order to relate the time of arrival of the wave at both ends.

- A refinement of this method is proposed in [80] that uses measurement data at one terminal alone. The arrival time of the first wave and those of succeeding reflections are used to estimate the fault location. With the help of simulation a fault location in the range of 3 to 5 % was estimated.

---

[1] A power outage during the months July till August 1996 affected millions of end customers in the south west of the United States and in northern Mexiko. More recently on Aug. 14th 2003 a power outage affected the entire north-east region of the United States and Canada including cities like New York City, Detroit, Ottawa and Toronto.

[2] A power outage affected the end customers in Great Britain (Aug. 29, 2003) and a second one those in Italy (Sept. 28, 2003).

[3] In February 1998 large districts of Auckland experienced a major power outage for several weeks after four central power cables broke down.

Off-Line fault location techniques:

Pulse methods, see [64] for an overview, are still generally used for fault location in underground cable systems when a fault is permanent. Depending on the kind of fault (high- or low-resistance, shunt or series fault, etc.) one of the following methods is usually selected:

- In the decay method, the voltage source has a high impedance in series with it and the voltage transient in the cable is measured. A high voltage is applied to the cable, inducing a breakdown at the fault. A transient is generated which travels back and forth between the fault site and the voltage source. The voltage transient is measured using a voltage-coupling device with a frequency response adequate to resolve both the edges and step portions of the voltage transient. The propagation of the transient is used to determine the fault location.

- In the current impulse method a surge generator applies a high-voltage step to the cable under test that induces a breakdown in the cable. The transient travels back and forth between the surge generator and the fault. The current transient is measured using a current transformer with a frequency response adequate to resolve only the edges of the current transient; this in turn is related to the fault distance, see [26].

- Time domain reflectometry takes advantage of the fact that impulses are reflected to some extent at cable discontinuities. When the impedance at the fault location approximates 0 or ∞ a distinctive reflection will occur and thus allow for a good fault location. One flaw of this method is that high-impedance shunt faults are difficult to localize.

- Arc reflectometry enhances the time domain reflectometry method by temporarily converting a high-resistance fault into a short circuit. This is facilitated by applying a high-voltage surge to the cable that ignites an arc at the fault location. Time domain reflectometry is then used to determine the fault location.

- After the approximate position of the fault has been found using one of the above methods, some means of pinpointing the fault must be used. Common practice has been to apply repeated high-voltage surges to the cable and listen for the "thump". To date, this acoustic method is by far the most successful technique used to pinpoint the precise location of a fault. However, indiscriminate use of a surge generator can place sound insulation at risk.

The off-line fault location techniques make it possible to pinpoint cable faults to within a few meters. In order to accomplish this satisfying result the defect cable needs to be isolated from the power grid and a high voltage generator next to the measurement equipment must be connected.

In order to implement an on-line method that delivers exact fault location one could use a tight synchronized distributed system (c.f. [8] or [20]) combined with the proposed techniques in [80] and [42] in order to detect and timestamp the first wave emanated after a surge stresses a cable. This in turn would decrease the time required for fault location and could thus reduce the impacts impaired by harmful calamities.

A similar application domain is on-line monitoring of partial discharges emitted in the cables of power distribution systems. Cavities in the isolation cause small surges that induce small travelling waves superimposed onto the supply voltage. As with fault-location one could monitor these effects and deduce potential problems that may arise due to electrical stress or aging. This information in turn is useful for a maintenance schedule of the employed cabling. These and several other uses of synchronized clocks in power distribution systems have been proposed to increase the quality of service, see [28].

Although these are by now only some few practical applications that could benefit from a very tight clock synchronization, others in computer science and the area of measurement and instrumentation will eventually emerge when this enabling technology becomes widely available. In particular applications like multimedia, mobile computing or process control applications, e.g. in papermills, will benefit from this service.

1.3 Outline

This thesis is structured as follows:

Chapter 2 proposes a taxonomy adapted to several clock synchronization algorithms: deterministic and probabilistic, internal and external. This taxonomy makes it possible to classify existing clock synchronization algorithms according to their internal structure and several basic building blocks. The following analysis is used to identify the building blocks that require support by hardware in order to provide tighter synchronization. Furthermore, the key parameters are developed out of existing algorithms that will set the physical limits for the achievable tightness.

Chapter 3 differentiates related work under the prospect of these limiting parameters. We briefly illustrate the hardware support for clock synchronization as used in the MARS project and by our *Network Time Interface* (NTI) M-Module and depict the shortcomings of these approaches. Next we address clock synchronization as used in the time triggered architectures TTP and FlexRay. In these systems clock synchronization plays a fundamental role for future time-critical avionic and automotive systems. Finally, the new IEEE Standard 1588 is considered and analyzed under these terms.

Chapter 4 introduces and discusses several hardware architectures to enable tight clock synchronization in modern popular switched Ethernet networks. In particular, network interfaces for end-systems and modifications of existing switch architectures are presented. The proposed approaches are analyzed under the prospects of how to achieve tight synchronization and how to keep modifications and influences on existing network hard+software low. Furthermore, this chapter briefly addresses enhancements for an integrated clock circuit, based on our UTCSU Asic, that is currently developed within the PSynUTC FIT-IT[4] project by the spin-off company Oregano Systems[5].

Chapter 5 describes the measurement setup used for an experimental evaluation of the underlying parameters that limit the achievable clock precision. In particular, we address the jitter of the physical layer of switched Ethernet networks under various different end-to-end configurations. Next, we briefly illustrate the analysis process of the gathered measurement data before we finally present the results obtained from the conducted experiments.

A short summary concludes this thesis, providing directions for future related research

---

[4]http://www.fit-it.at
[5]http://www.oregano.at

issues. In particular, we give a summary of extensions to our approach in order to evolve an implementation that can lead to an industrial realization.

Chapter 2
State of the Art of Clock Synchronization

In centralized systems, mutual exclusion and inter task communication problems are generally solved using methods such as semaphores and monitors and highly rely on shared memory. This is not true for distributed systems where even the simplest thing such as determining whether event A happened before or after event B requires careful thought. In general, distributed systems have the following characteristics [72]:

- Multiple concurrent computation threads,
- interconnections for inter-thread communication and a
- global, shared state the individual computers cooperatively maintain.

To achieve a consistent global state it is necessary to address the issues of independent node failures and unreliable, insecure and costly communication. To that end, clock synchronization provides internal consistency of the clocks at the distributed nodes that eases the implementation of the before-mentioned issues. During the past few years, much research has been conducted towards a common view of time in fault-tolerant distributed systems. There are more than 60 papers listed in a 1993 bibliography [114] on clock synchronization in distributed systems. As a result, the proposed algorithms are relatively well understood. The increasing demand for ever tighter synchronization and some additional requirements, e.g., to synchronize with an external time standard, resulted in further research, see [86] for an overview.

This thesis proposes a mechanism for further enhancements of existing clock synchronization strategies. The presented approach exploits the limiting parameters of existing implementations. Therefore, a comparative study and an in-depth analysis of published algorithms seems appropriate.

Software clock synchronization algorithms use standard communication networks and send synchronization messages to get the clocks synchronized. They are more frequently used since a loose synchronization in the range of some *ms* is acceptable in most applications. All software clock synchronization algorithms proposed so far decompose themselves in *deterministic*, *probabilistic* and *statistical* algorithms. Deterministic algorithms, e.g. [78, 16, 24, 25, 27, 56, 63, 85, 87] assume an upper bound on transmission delays and guarantee a maximum difference between any two simultaneous clock readings. Probabilistic algorithms [2, 75, 18, 15, 78] guarantee a constant maximum deviation between synchronized clocks. In particular, a clock knows at anytime if it is synchronized or not with the other, but there is a non-zero probability that a clock will get out of synchronization when too many unmasked communication failures occur. Statistical algorithms

[113, 14, 58] assume that the expectation and standard deviation of the delay distributions are known. Clocks do not know how far apart they are from each others, but a statistical argument is made that at any time, any two clocks are within some constant maximum deviation with a certain probability.

Hybrid clock synchronization [53, 34] solutions based on software algorithms with moderate hardware support achieve reasonably tight synchronization and are still cost-effective in comparison to pure hardware approaches. The embodied hardware support usually maintains the local clock, applies the required corrections and provides some facilities to ease the exchange of clock messages. The proposed clock synchronization method fits into this hybrid category and extends related work by providing a new way of incorporating clock messages into state-of-the-art network solutions.

Pure hardware-based clock synchronization [99, 106, 12] achieves very tight synchronization through the use of special synchronization hardware at each processor, and uses a separate network solely for clock signals. Due to cost, size and practicality reasons the additional network is in most cases only affordable when the system spread is within the range of a few meters. The primary application domain of this kind of synchronization strategy are parallel systems where a set of microprocessors needs to coordinate their actions. This thesis does not consider these specialized, costly mechanisms, instead it concentrates and exploits mechanisms of software based and hybrid clock synchronization.

The remainder of this chapter provides a classification and analysis of published clock synchronization algorithms. The aim herein is to devise a good understanding of the principles involved in order to extract the relevant parameters one needs to tackle to achieve tight synchronization. Similar surveys, but with a more general focus in mind, can be found in [93, 82, 100] and [1]. In [82], software and hardware clock synchronization algorithms are classified with regard to the clock correction scheme used. In contrast, the algorithms surveyed in [100] are listed according to the supported faults and the system synchrony (knowledge of upper bounds on communication latencies). A very thorough classification is given in [1] that aims to help the designer in choosing the most appropriate structure of algorithm and the best building blocks suited to his/her hardware architecture, failure model, quality of synchronized clocks and message cost induced.

A short section on system modelling defines relevant parameters and properties as used in most relevant papers on this topic. The following taxonomy tries to figure out the delimiting parameters for an achievable clock synchronization tightness. The results of this analysis are used to identify requirements and improvements for hardware support for tight clock synchronization.

2.1 System Modelling

A comparative study of the impaired parameters of clock synchronization algorithms requires establishing a general system model. According to the analyzed algorithms we consider a set of distributed nodes interconnected by a communication network that can have different characteristics (broadcast or point-to-point, fully-connected or not). Every node hosts a *Central Processing Unit*, some kind of *Communications Controller*, a storage device in the form of a local memory and a local clock. In order to allow for a synchronization with real-time some nodes should be equipped with GPS (Global Positioning System) or other reference timing receivers in order to obtain the time signal broadcast by a standard source of time, as UTC (Universal Time Coordinated) the official time standard.

2.1.1 Clocks and Processors

For an arbitrary node p, clock C_p generally consists of an oscillator O_p and a counting register that is incremented by the ticks of the oscillator. While clocks are discrete, having non-zero granularity G, all algorithms assume that clocks run continuously, i.e., C_p is assumed to be a monotonic real-valued function of real-time t.

Clock Precision: The paramount problem of distributed clock synchronization is to maintain the maximum clock state deviation between any two clocks at different correct nodes p and q at all real-times bounded by a value called *precision* π

$$|C_p(t) - C_q(t)| \leq \pi \quad \forall t \in T.$$

This problem referred to as *internal clock state synchronization* is usually at the core of all clock synchronization algorithms.

Accuracy: Synchronizing the clock C_p of one arbitrary node p of a distributed system with an external time standard, e.g. UTC, is denoted *external clock state synchronization*. It keeps the maximum deviation between corresponding clock states and real-times on a single correct clock bounded by a constant called *accuracy* α

$$|C_p(t) - t| \leq \alpha_p \quad \forall t \in T.$$

Accuracy becomes essential when the spatial diameter of the distributed system is very large or when the system needs to interact with other systems respectively. In both cases a common notion of time that is synchronized closely to the official time standard becomes crucial.

Clock Drift: It is commonly assumed that even correct oscillators exhibit some instabilities due to temperature changes, aging and other reasons. In general, the *oscillator drift* ρ is the systematic change in frequency with time of an oscillator. The manufacturer usually specifies a maximum ρ and hence bounds the instantaneous oscillator frequency $f(t)$ by

$$(1-\rho) \leq \frac{f(t)}{f} \leq (1+\rho).$$

Direct coupling of an oscillator with the clock yields a clock rate of $Sf(t)$ where the coupling factor S is the constant $1/f$, i.e. $C_p(t) = \frac{1}{f}\int_0^t f(t)dt$. The maximum deviation between the clock rate and the ideal rate 1 is denoted by the *clock drift* ρ_p, formally

$$\left|\frac{dC_p(t)}{dt} - 1\right| \leq \rho_p \quad \forall t \in T.$$

Thus a perfect clock has $dC/dt = 1$, a fast clock $dC/dt > 1$ and a slow clock $dC/dt < 1$. If two clocks are drifting in opposite direction, at a time Δt after they were synchronized, they may be as much as $2\rho\Delta t$ apart. Thus, in order to guarantee that no two clocks ever differ by more than Δt, clocks must be synchronized at least every $\Delta t/2\rho$ seconds.

Consonance: The maximum clock rate deviation between two different correct clocks in the distributed system at simultaneous real times is called *consonance* γ [95, 96], formally

$$|\frac{dC_p(t)}{dt} - \frac{dC_q(t)}{dt}| \leq \gamma \quad \forall t \in T.$$

Maintaining consonance resp. drift of an ensemble of clocks refers to the problem of *external* resp. *internal clock rate synchronization* [96].

Initial synchronization: A sometimes neglected aspect is the problem of system start-up and node join and how initial synchronization is accomplished respectively. An assumption made by several algorithms makes it necessary that the nodes clocks are initially synchronized and that this initial synchronization is bounded by a given constant β, see [63] for example.

The clock synchronization algorithm executes on the local processor and takes the clock readings of the local and remote clocks as inputs. The required computation time should be small and bounded by some constant value in order to account for clock state and rate changes during this time span. The computed correction is afterwards applied to correct the local clock time. Most systems are equipped with a pure oscillator+counter based clock where synchronization of the local hardware clock is not possible at all. Rather, logical clocks are introduced. The value of a logical clock at real-time t is determined by adding an adjustment term to the local hardware clock $C_p(t)$. The adjustment term can be either a discrete value, changed at each re-synchronization [92, 101], or a linear function of time [92, 15, 91]. The discrete clock adjustment technique may cause a logical clock to instantaneously leap forward or be set back, and then continue to run at the speed of its underlying hardware clock. Such behavior cannot be tolerated by most distributed applications requiring clock synchronization, therefore a linear function of time for clock adjustment is often mandatory.

2.1.2 Communication Subsystem

In distributed real-time systems, message delays may be more or less predictable depending on the type of network used and the assumptions made on the network load. Some algorithms assume that known lower and upper bounds to deliver (i.e. to send, transport and receive) a message between correct nodes exist. The mechanism to send, transport, and receive any message over a correct link from a correct node p to node q experiences a delay $\triangle t'_{p,q}$ subject to the delay condition

$$\triangle t_{p,q} - \varepsilon_{p,q} \leq \triangle t'_{p,q} \leq \triangle t_{p,q} + \varepsilon_{p,q},$$

where $\triangle t_{p,q}$ represents the deterministic part and $\varepsilon_{p,q}$ the delivery uncertainty with $\triangle t_{p,q} \geq \varepsilon_{p,q}$. When this assumption holds, a deterministic clock synchronization algorithm can ensure that all correct logical clocks are within a maximum distance from each other.

In practice from the mechanisms involved in present communication subsystems one main limiting factor for clock precision and accuracy is ε. In fact, the work of [62] revealed that even n ideal clocks cannot be synchronized with a worst case precision less than

$$\varepsilon(1 - \frac{1}{n}) \tag{2.1}$$

in presence of a delivery uncertainty ε. Unfortunately, in a shared channel type network there are several steps involved in packet transmission/reception that could contribute to ε, cf. [53] and [34, 89]. The following general steps illustrate a usual message transfer from node p to node q:

(1) The CPU at node p assembles the packet (including a local timestamp) and notifies the Communication Subsystem.

(2) The Communication Subsystem at node p in turn acquires the network medium and sends the resulting bit-stream.

(3) The receiving communications module at node q pulls the bit-stream from the medium and notifies the CPU at node q via interrupt of the current packet reception.

(4) The CPU at node q processes the received packet and marks the reception time by reading its local clock.

Although this scenario is by far idealized, it becomes apparent that ε is affected by several sources of indeterminism. The medium access uncertainty at (2) strongly depends on the access policy of the used bus system. The network delay $(2 \rightarrow 3)$ may vary due to queuing delays at intermediate gateway nodes. The reception interrupt latency until the CPU processes the packet $(3 \rightarrow 4)$ strongly depends on the CPU load.

2.1.3 Faults

For development and analysis of distributed algorithms faults need to be considered since they may affect every single component (processor, communication links, clocks, etc.) of the system. A proper fault model $\mathcal{F}$ needs to be specified in order to elaborate on system operating conditions when faults occur. Following is a list of types of processor, link and clock failures that have mostly been assumed throughout clock synchronization algorithms.

Concerning clocks, most of the algorithms assume uncontrolled failures (also called Byzantine or arbitrary failures). Other ones assume timing failures, a more restricted failure mode prohibiting conflicting information.

Clock Byzantine failure: A local hardware clock commits a Byzantine failure when it gives inaccurate, untimely or conflicting information. This includes dual-faced clocks, which may give different values of time to different processors at the same real-time.

Clock timing failure: A local hardware clock commits a timing failure if it does not meet the clock drift condition, i.e., is not ρ-bounded.

The failure semantics of processors assumed in published algorithms cover nearly all the kinds of failures ever identified. Processors may crash, commit performance failures, or more generally, commit Byzantine failures.

Processor crash failure: A processor commits a crash failure if it behaves correctly and then stops executing forever (permanent failure).

Processor performance failure: A processor commits a performance failure if it completes a step in more than the specified time.

Processor Byzantine failure: A processor commits a Byzantine failure if it returns incorrect or malicious data, see [57].

With regard to the communications subsystem, whatever its type (broadcast or point-to-point), the failure semantics are more restricted. A link may commit omission or performance failures but must never partition the network.

Link omission failure: A link from a node p to a node q commits an omission failure on a message if the message is inserted into p's outgoing buffer but the link does not transport it into q's incoming buffer.

Link performance failure: A link commits a performance failure if it transports some message in more time than specified. Clearly, this applies only to systems with known upper and lower bounds on transmission delays.

2.2 A taxonomy of clock synchronization algorithms

Having defined several parameters required for system modelling, this section is devoted to identifying common building blocks of clock synchronization algorithms. The proposed taxonomy relies on two orthogonal features: the internal structure and the basic building blocks from which most clock algorithms are built and thus follows and extends the presentation given in [1]. The internal structure represents the way synchronization messages are distributed and the role certain nodes play in synchronization. The building blocks correspond to successive steps executed by every clock synchronization algorithm and are kept generic in the sense that they apply to several different kinds of algorithms — deterministic, probabilistic and statistical; internal and external.

2.2.1 Structure of clock synchronization algorithms

The structure of clock synchronization algorithms is discerned by the way how time is disseminated and how every node participates in the clock synchronization.

Asymmetric (Master-Slave) Structures

Throughout the internet time-servers are deployed and client nodes may synchronize to the time provided by these servers using the *Network Time Protocol*. A similar *master-slave* structure is employed by other synchronization strategies as described in [27, 15, 2] and may be classified as an *asymmetric* scheme. Usually one dedicated node is designed as master and provides the time to the other nodes designed as slaves. In some implementations, denoted as master-controlled schemes, the master acts as coordinator of the clock synchronization algorithm. It requests and collects the slaves clocks, estimates the required adjustments and sends back the corrected clock values. On the other hand, in slave-controlled schemes, the master provides only the reference time. Every slave asks for the reference time and after reception invokes the clock synchronization algorithm and re-synchronizes the local clock. The advantage of asymmetric schemes is their low cost in terms of number of messages exchanged. On the other hand, the most common drawback is given by the presence of the master, which represents a single point of failure. In addition, a single master can be swamped by a large numbers of synchronization messages, thus invalidating in some way communication delay assumptions. To overcome this problem some extra mechanisms such as fault detection followed by the election of a new master, or duplication of masters are required.

Symmetric Structures

In *symmetric* schemes every node that participates in an active manner, executes the whole clock synchronization algorithm. Therefore every node disseminates its local clock value to all other nodes and, in turn, gathers the clock value from them. The received clock values in relation to the local clock are then used to compute a correction value. This adjustment term is afterwards applied to the local clock. Symmetric algorithms can be split in two classes, flooding-based and ring-based, depending on the virtual path taken for transmitting a message from one processor to every other one. In flooding-based symmetric algorithms (see e.g. [108, 16, 101, 63, 79]), each processor sends its messages to all outgoing links. Messages received on incoming links are relayed when a non-fully-connected network is used. The benefit of flooding-based techniques is their natural support for fault tolerance, i.e., they do not exhibit a single point of failure. However, they may require up to n^2 messages to disseminate a message to all nodes in the system, with n being the number of nodes in the system. This large number of messages can be lowered to n if a broadcast network is used. The virtual ring scheme was designed in order to decrease the number of exchanged messages experienced in the flooding-based schemes, cf. [74]. In the ring scheme, all the processors in the system are gathered along a virtual ring. The number of messages is reduced by sending only one message along this cyclic path. As this message travels on the ring, each processor adds its own data to the message. Compared with flooding-based schemes, virtual ring schemes need a smaller number of message exchanges (only n messages per re-synchronization are used), but need extensions in order to support node failures.

Hierarchical Structures

Some clock synchronization algorithms consider a *hierarchical* clock synchronization strategy where the synchronization is spread at different levels. The operation at one distinct level may again be categorized either as asymmetric or symmetric. In addition, nodes may participate in the clock synchronization in a different manner. Passive nodes may synchronize their clocks but may not contribute to the overall synchronization in contrast to active nodes that will provide their time for other nodes as well. Primary nodes are either equipped with a better oscillator or may have access to an external time-source, e.g., a GPS receiver in contrast to secondary nodes. In general, not all clock synchronization algorithms impose a structure. This way they can be employed in both either asymmetric or symmetric schemes.

2.2.2 Clock synchronization building blocks

Clock synchronization in a distributed system is composed of several fundamental building blocks:

- Re-Synchronization event detection
- Remote clock estimation technique
- Clock correction block

Re-Synchronization event detection block

A clock synchronization algorithm has to detect the instant at which it must re-synchronize. Due to clock drift, clocks must be re-synchronized frequently to guarantee precision π and accuracy α. Usually clock synchronization algorithms are round-based, each round being devoted to the re-synchronization of all clocks. Thus this *re-synchronization event detection block* becomes active periodically. The difficulty arises when dealing with the time at which rounds must start. One technique assumes initially approximately β-synchronized clocks, see e.g. [63, 56, 79, 91]. Some external means are usually required to provide this initial condition. Ensuring this setting a node considers the start of a new round k when its local clock reaches kP, where P is the round duration in local time, i.e., the time between two successive re-synchronization rounds. Intuitively, to keep clocks as closely synchronized as possible, β and P must be as small as possible. However, P cannot be arbitrarily small in order for any algorithm to work correctly, cf. [63] or [91]. Another technique uses message exchanges to invoke a synchronization round, see [101, 108] for examples. A node sends a message to all other nodes when its local clock reaches a predefined value. It starts a new synchronization round upon reception of a fixed number of such messages originating from other nodes. The number of expected messages depends on the maximum number of failures assumed. The message latencies thus directly influence the achievable precision, since rounds are triggered on message reception. By using broadcast networks, exhibiting a small variance of transmission delays, precision can be improved [108].

Remote clock estimation technique

When a new clock synchronization round initializes, every node tries to get some knowledge of the value of remote clocks. Due to variable communication delays and clock drifts only estimates can be acquired. It is essential for any clock synchronization strategy, that these estimates are closely related to the remote clock values since the clock readings form the input for the subsequent clock correction block. The *remote clock estimation technique* operates as follows: Each node sends its local clock value T encapsulated within a message to every other node. The receiving node uses the message contained therein to estimate the clock of the sender. This is possible when communication delays are bounded and clocks are initially synchronized. The receiver feeds the remote clock values and the corresponding local times at which messages are received to the clock correction block. Only clock messages received within an interval of length $(1+\rho)(\triangle t_{p,q} + \varepsilon_{p,q} + \beta)$ following the most recent re-synchronization event are considered. The value of the remote clock at node q belongs to the interval

$$[T + (1-\rho)(\triangle t_{p,q} - \varepsilon_{p,q} - \beta), T + (1+\rho)(\triangle t_{p,q} + \varepsilon_{p,q} + \beta)]$$

seen at node p, cf. [63]. When ignoring smaller order terms, the maximum difference of two remote estimates of different nodes when communication delays are bounded is given by

$$2(\varepsilon_{p,q} + \beta + \rho \triangle t_{p,q}).$$

In the case where communication delays are not bounded and clocks are not initially synchronized a set of successive transmissions of timestamped synchronization messages are at hand. The clock correction block uses message delay statistics (expectation and deviation of transmission delays) for computation of a correction term.

A remote clock estimation mechanism that can be used in the absence of an upper bound on communication delays has been introduced in [15]. Cristian's master-slave algorithm is based on a remote clock reading technique that allows estimating a remote clock to lie within a given interval and is used to achieve external clock synchronization. In order to obtain a remote clock reading from some node q a process p sends a

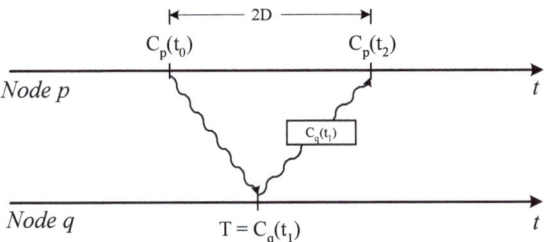

Figure 2.1: Remote Clock Reading

requesting message to node q at $C_p(t_0)$ querying q's time. The remote node replies with a message that encapsulates its clock value $T = C_q(t_1)$. This message, in turn, is received at node p at its own local time $C_p(t_2)$. With the knowledge of the round trip delay $2D$ node p can estimate q's clock and bound the error it makes when reading q's clock. Since p's clock can drift from real-time by at most ρ, the round trip delay is not greater than $(C_p(t_2) - C_p(t_0))(1+\rho) = 2D(1+\rho)$. The one-way transmission delay for the reply message may hence be defined by a constant $max := 2D(1+\rho) - min$ where $min = \triangle t_{p,q} - \varepsilon_{p,q}$ accounts for the minimum transmission delay. These two bounds can now be used to approximate q's clock at time $C_p(t_2)$ since q's clock value increases by $min(1-\rho)$ at the least and by $max(1+\rho)$ at the most, respectively. Hence q's clock value T estimated at node p at $C_p(t_2)$ lies in the interval:

$$C_q^p(t_2) \in [T + min(1-\rho), T + max(1+\rho)]$$

Since node p has no means of knowing exactly where the clock of node q lies within the above interval it estimates the value with a function $C_q^p(T,D)$ that for e.g. chooses the midpoint of this interval. The inherent maximum error made therein is given by

$$2D(1+\rho)^2 - 2(\triangle t_{p,q} - \varepsilon_{p,q}) \sim 2D(1+2\rho) - 2(\triangle t_{p,q} - \varepsilon_{p,q}).$$

If node p wants to achieve a reading error smaller than a certain specified maximum error, it must discard any reading attempt for which it measures an actual round trip delay $2D > 2U$ with $2U$ denoting the *timeout delay* necessary for achieving the required reading precision.

Clock correction block

The local and remote clock readings serve as input to a suitable clock correction algorithm. The latter calculates clock correction terms that are applied succinctly to the clocks. Algorithms presented in scientific literature can be categorized into several categories:

Clock State Correction: A clock at every node maintains local time and hence allows us to tell when a particular event occurs or how long a duration takes. Most clock synchronization algorithms in general, and clock correction functions in particular, aim at keeping clock states together as well as the deviation towards real-time bounded. Viewed under this aspect most algorithms presented in scientific literature can be categorized into

- internal
- and internal and external clock synchronization.

Further categorization is due to the implementation of the clock correction algorithm as

- deterministic,
- probabilistic and
- statistical.

Clock Rate Correction: Clock synchronization can be viewed either in terms of clock speeds or by considering the instantaneous clock rate $v(t) = dC(t)/dt$. The goals here are to keep the deviation between the clock rate and the ideal rate 1 and the clock rates between two different nodes at simultaneous real-times bounded.

The remainder of this subsection informally describes the principles of a selected set of clock synchronization algorithms and lists the precision and accuracy they achieve[1]. The aim herein is to give a short introduction and to allow extraction of a set of parameters that need improvements through appropriate hardware support.

Deterministic clock synchronization:
Most clock correction blocks implemented in round based algorithms use the notion of a *convergence function* ($C\mathcal{V}$) introduced in [94]. These algorithms can be generically described as follows: At the end of a synchronization round every process reads the clocks of all processes and then adjusts its clock value for the next round by applying a convergence function to the clock readings of the current round. A synchronization algorithm that can be obtained from the above generic algorithm by instantiating some concrete function for the abstract notion of a convergence function will be termed a *convergence function based algorithm*. Convergence functions use the set of remote clock estimates to compute a new clock value. For this, special averaging techniques are at hand; in addition, they usually provide a mechanism for tolerating erroneous clock readings as well. The most prominent and promising convergence functions are briefly listed below along with a short informal description.

$C\mathcal{V}$-functions for internal clock synchronization:

- In the interactive convergence algorithm, see [56], each process reads the value of every process's clock and sets its own clock to the average of these values — except when it reads a clock value differing from its own by more than $\triangle \approx \pi + \varepsilon$,

---
[1]The symbols and notation used for several parameters follows the definitions given in Sec. 2.1 rather than those of the literature, to ease comparisons and reasoning.

then it replaces that value by its own clock's value when forming the average. The achievable worst case precision of this algorithm is given by:

$$max(\frac{n}{n-3m}(2\varepsilon + \rho(P + 2\frac{(n-m)S}{n})), \beta + \rho P)$$

with S denoting the final seconds of the interval P, n being the total number of processes and m the number of faulty ones.

- The master-slave clock synchronization implemented in TEMPO, the distributed service that synchronizes the clocks of Berkeley UNIX 4.3 BSD, is described in [27]. A master time daemon measures the time difference between the clock of the machine on which it runs as well as those of all other machines. The master computes the network time as the average of the times provided by non-faulty clocks. A clock is considered faulty if its value is more than a small specified interval away from the values of the clocks of the majority of the other machines. The master then sends to all slave time demons —also to those with faulty clocks— the correction that should be performed on the clock of its machine. Since the correction can be negative, in order to preserve the monotonicity, TEMPO implements the correction by slowing down (or speeding up) the clock rates.

  When the master time demon synchronizes every P seconds all non-faulty clocks are within range:
  $$4(D - 2min(T_{pq}^{min}, T_{qp}^{min})) + 2\rho P$$
  with T_{pq}^{min} and T_{qp}^{min} denoting the minimal possible transmission times from nodes p to q and from q to p respectively.

- The *fault-tolerant midpoint function* proposed in [63] synchronizes an ensemble of $3F + 1$ nodes where at most F of these are faulty. The algorithm executes in a series of rounds; each round is started when a clock reaches a certain predefined value. Meanwhile, for a bounded amount of time, it collects clock messages from all other nodes. Then every node invokes the fault tolerant midpoint function that returns the midpoint of the range of clock values received from all other nodes after the f highest and f lowest values have been discarded: $FTM(p, \theta) := mid(sort(\theta)(F), sort(\theta)(N - F - 1))$. An analysis of the algorithm presented in [24] shows that the maximum deviation between correct clocks is $4.5\Lambda + 4\rho r_{max}$ considering initial synchrony, with 2Λ accounting for the clock reading error, ρ the clock drift and r_{max} the maximum (real-time) duration of a round.

In addition, Fetzer and Cristian derived lower bounds for convergence function based clock synchronization algorithms [23]. In particular, the lower bound for the maximum deviation is given by
$$4\Lambda + 4\rho P$$
Given this lower bound they proposed a convergence function termed *differential fault tolerant midpoint convergence function* that guarantees an optimal correction, an optimal maximum drift rate, and optimal deviation.

- The differential fault-tolerant midpoint function is based, as its name implies, on the fault-tolerant midpoint function. First, it defines an *extended midpoint function* by computing the midpoint of the union of the intervals $[Y(F), Y(N-F-1)]$ and $[T-\varepsilon, T+\varepsilon]$ for a given time T and an array Y of N clock values:

$$emid := mid(min\{T-\varepsilon, Y(F)\}, max\{T+\varepsilon, Y(N-F-1)\})$$

The differential fault-tolerant midpoint function calculates the new clock value for a given process p and a clock reading Θ, where p's own clock shows $\Theta(p)$ at the end of the round and the function *sort* returns a sorted array of the clock values in Θ:

$DFTM(p, \Theta) :=$
if $|emid(\Theta(p), sort(\Theta)) - \Theta(p)| \leq 2\rho P$ **then**
 $emid(\Theta(p), sort(\Theta))$
else
 $T + sign(emid(\Theta(p), sort(\Theta)) - \Theta(p))2\rho P$
end if

This convergence function bounds the drift rate of correct clocks by ρ, the maximum correction by $2\rho P$ and the maximum deviation by $4\varepsilon + 4\rho P + 2\rho \beta$. Hence this algorithm is optimal with regard to the lower bounds given for convergence function based clock synchronization.

- The adaptive exponential averaging fault-tolerant midpoint function [3] was developed in order to avoid excessive clock corrections that are encountered with the original fault-tolerant midpoint function. Its operation is based on the principle that excessive clock correction terms are more likely caused by long message delays than by a transient fault in the node computing the clock correction term. An adaptive varying weighting factor relates the correction term used during the last resynchronization round to the new derived correction term computed with the fault-tolerant midpoint function.

 For not completely connected networks the same authors presented in [4] the multi-step interactive convergence function that extends the fault-tolerant midpoint function to hierarchical partitioned networks. Clock synchronization executes within m steps, each step is devoted to the synchronization of a set of nodes at a higher hierarchical level.

- The sliding window function proposed in [79] achieves an increased fault tolerance without the disadvantage of reduced synchronization tightness. At a given round the algorithm slides a window of fixed width $w \approx \pi + \varepsilon$ starting at the leftmost element over a sorted array of remote clock readings. A window instance is spanned whenever the left border of the window is aligned with a clock reading. Among all possible window instances the algorithm selects the instance that contains the maximum number of clock readings. If two or more instances exist that include the same number of clock readings then either the first instance is selected or the one with the minimum variance of included clock readings. The convergence function then calculates the mean or the median of all clock estimates within the given selected window instance, respectively. The best achievable precision is given by

$$\frac{(n^2+n-f^2)\varepsilon + 2P\rho(n-f)(n-f+1)}{n^2 - 5fn + n + 4f^2 - 2f}$$

for n nodes with maximum $f \leq \frac{n}{4}$ nodes exhibiting Byzantine behavior. For n=2 this results in approximately $\sim 5,75\varepsilon + 5P\rho$

$\mathcal{CV}$-functions for internal and external clock synchronization:

- In [25] the integration of fault-tolerant external and internal clock synchronization is addressed. The proposed algorithm provides both external and internal clock synchronization for as long as a majority of nodes that provide access to a reference time stay correct. In particular, the algorithm assumes that from a given set of $\geq 2F+1$ reference time servers at most F servers suffer arbitrary failures. Every non-reference time server periodically reads all reference clocks, sorts the readings and stores them in an increasing array. In this set, amongst the $F+1$ smallest clock readings, there exists at least one reading that belongs to a correct reference clock. The same applies to the $F+1$ greatest clock readings. Hence there is at least one intersecting clock reading that can be used for synchronization. This algorithm achieves a maximum external deviation of $\Delta+\epsilon+\rho P$ and a maximum internal deviation of $min\{2(\Delta+\epsilon+\rho P), 4\epsilon+9\rho P+2\rho\beta\}$ with Δ denoting the maximum external deviation of correct reference clocks.

  When the majority of nodes that access the reference time, or when the reference time itself becomes unavailable, the algorithm switches to a degraded mode where only internal clock synchronization is performed. In this state the maximum internal deviation is bounded by $4\epsilon+9\rho P+2\rho\beta$ and the external deviation becomes unbounded.

- The *orthogonal accuracy* convergence function $\mathcal{OA}$ presented in [87] operates on intervals where every interval is given by the clock value and a corresponding accuracy interval which implies the maximum deviation of the local clock value to an external reference time. This set of clock and accuracy information is maintained at every node that participates in the clock synchronization process and is exchanged periodically in rounds to all other processes. At the end of a round each node applies the $\mathcal{OA}$ function to the received interval set. Basically, the result of $\mathcal{OA}$ is the interval provided by the Marzullo[2] function $\mathcal{M}$ applied to the set of accuracy intervals and extended appropriately to include the reference point. The reference point is computed independently by $\mathcal{OA}$ as the *center* of an interval obtained by applying $\mathcal{M}$ to the associated *precision intervals* of an according input set. The precision intervals are related to an "artificial" internal global time which is given for each round and progresses as real-time does, see [87, 88].

  The comprehensive analysis given in [87], which relies on the interval-based framework established in [91], yields a maximum precision and clock correction of $5\epsilon+4P\rho$ and global drift $\rho+\epsilon/P$ plus some smaller terms. Thus, with respect to worst case precision $\mathcal{OA}$ performs equivalent to the fault-tolerant midpoint function. The very detailed system model used for the analysis of the algorithm properties revealed that — apart from clock reading error, clock drift, re-synchronization period and external deviation — the clock reading granularity G, the clock setting granularity G_s and rate adjustment uncertainty u have an impact on the achievable precision and accuracy (as much as $12u+4G+G_s$). This makes it clear that any

---

[2]$\mathcal{M}$ is a fault-tolerant intersection function that was introduced in Marzullo's thesis [65] and termed *Marzullo function* in [55]. $\mathcal{M}_n^{n-f}(I)$ is defined as the largest interval whose edges lie in the intersection of at least $n-f$ different intervals I_j for a given set of compatible intervals $I = I_1, \ldots, I_n$ with $n \geq 1$ and $n-f \geq 1$. In other words, the resulting interval of $\mathcal{M}_n^{n-f}(I)$ is determined by sweeping from left or right to the center over the set of interval endpoints and stopping at the $n-f$ endpoint, the latter defining the left and right endpoint of the resulting interval, respectively.

attempt to improve on the worst case precision and accuracy must pay attention to these parameters as well.

- The *optimal precision* convergence function $\mathcal{OP}$ [88] enhances the $\mathcal{OA}$ function by relating the local accuracy interval and clock value to the set of remote information. The corrected accuracy interval at an arbitrary node q is obtained by forming the intersection $\mathcal{M}(I_q) \cap I_q^q$, where $\mathcal{M}(I_q)$ is the Marzullo function that is applied to the set of remote interval readings and I_q^q denotes the accuracy interval originating in the node's own clock. In a similar way, the reference point is set to the *center* of the intersection given by the *precision interval* from the local node q and the Marzullo function applied to the remote *precision intervals*. Algorithm $\mathcal{OP}$ yields optimal worst case performance given by the maximum precision $4\varepsilon + 4P\rho$, the maximum clock correction $2P\rho$ and global rate ρ. The worst case accuracy intervals can be as large as $3P\rho$. The same thorough analysis, as applied to algorithm $\mathcal{OA}$, revealed that the clock granularity G and the rate adjustment uncertainty u may have a considerable impact by as much as $11u + 3G + G_s$.

- A global time service for world-wide systems, dubbed *Cesiumspray*, is presented in [108]. The clock synchronization scheme employed is a hierarchical mix of external and internal synchronization where the GPS satellites form the root of the hierarchy which *spray* their reference time over a set of nodes provided with GPS receivers, one per local network. The second level of the hierarchy performs internal synchronization, further *spraying* the external time inside the local network. The algorithm of the second level is derived from the *a posteriori agreement* synchronization algorithm [107], modified to follow an external clock, and able to use simple group communication and membership facilities. The algorithm uses periodic broadcasts to disseminate clock values and ensures that all correct processors choose the same broadcast and adjustment to synchronize their clocks. Since the broadcast reception time is practically the same everywhere in the local area network it is possible to drastically attenuate the traditional limitation imposed by the message delivery delay variance on the obtained precision.

Probabilistic clock synchronization:
Most clock synchronization algorithms proposed in the literature try to guarantee an upper bound on the clock skew with certainty, cf. previous discussion. However, a theoretical limit is given by Equ. 2.1.

Clock skews that are significantly smaller than the theoretical limit that can be achieved if the requirement of determinism is relaxed and a probabilistic guarantee is accepted.

- In [15] the idea of probabilistic clock synchronization was proposed for the presence of unbounded communication delays. With the help of a modified remote clock estimation technique —see Sec. 2.2.2— this paper analyzes a master-slave clock synchronization mechanism. For at most k reading attempts during which the slave can drift from the master by as much as $\rho k(1+\rho)W$, Cristian derived a maximum external deviation that is given by $U - min + \rho k(1+\rho)W$ with $(1+\rho)W$ specifying the maximum real time which can elapse between successive reading attempts. For an aggressive setting where U is chosen close to $t - \varepsilon$ a maximum deviation as small as $\rho k(1+\rho)W$ can be achieved at the expense of many synchronization messages. U and W are constants chosen in a way to satisfy $2D < 2U < W$, where 2D is the measured round-trip delay.

- The probabilistic internal clock synchronization algorithm presented in [17] uses an improved remote clock reading method compared to [15]. In particular it exploits broadcast messages whenever possible to reduce the number of messages exchanged, it selects from all message pairs between two processes the message pair that provides approximately the lowest upper bound for the clock reading error and the processes stagger the sending of messages in time, to reduce network congestion and message concurrency. Four different clock synchronization algorithms are presented differing in the underlying failure assumptions. All four algorithms yield a maximum deviation between correct clocks of:

$$4\varepsilon + 4\rho r_{max} + 2\rho\beta$$

with ε, ρ, r_{max} and β denoting the clock reading error, the clock drift, the maximum real-time duration of a round and the maximum initial difference between two nodes.

- The probabilistic clock synchronization algorithm given in [2] relies on a master-slave variant of the *time transmission protocol* (TTP). When a node p wants to communicate its clock to a target node q, it sends a sequence of n synchronization messages encapsulating its own local time T. The target node q records the time R, according to its local clock, at which it receives each message. After receipt of n messages node q estimates the time on p's clock by using the following equations:

$$T_{est} = R_n - \frac{1}{n}\sum_{i=1}^{n} R_i + \frac{1}{n}\sum_{i=1}^{n} T_i + \overline{d}$$

with $\overline{d}$ estimating the expected value of the message delay. The maximum clock skew between any two clocks in the system that can be obtained by this algorithm is given by $\gamma_{max} = 2(\varepsilon_{max} + (R_{synch} + d_{max} - d_{min})\rho)$ where ε_{max} denotes the desired maximum skew at re-synchronization, R_{synch} is the specified re-synchronization interval, d_{max} and d_{min} are the maximum and minimum message delays and ρ is the relative clock drift. The author stated typical values for γ_{max} of about $2ms$ for a setting where in comparison deterministic algorithms could only achieve about $50ms$ given the theoretical bound from [62].

Statistical clock synchronization:
Statistical clock synchronization algorithms rely on a statistical learning of clock parameters, see [21], [44] or [14]. As a consequence, the precision and accuracy reached are generally very good even if transmission delays have a rather great dispersion, e.g. in the internet. The main drawback herein is that precision and accuracy cannot be specified in deterministic terms — instead quantile and confidence intervals are used therefore. These algorithms assume that all messages from a node p to node q are timestamped and that the transmission delay is random variable where neither its expectation nor its distribution are known. The algorithms in [21] and [14] apply a linear regression for a sequence of timestamped messages in order to compute the relative drift between the nodes. For computation of the initial offset a remote clock reading method as proposed in [15] can be employed.

In [113] a statistical method for time synchronization of computer clocks with precisely frequency-synchronized oscillators is proposed. A similar method is used in [58] to perform time synchronization using the internet. The most prominent and widely used synchronization mechanism is the *network time protocol* (NTP) proposed in [67] and

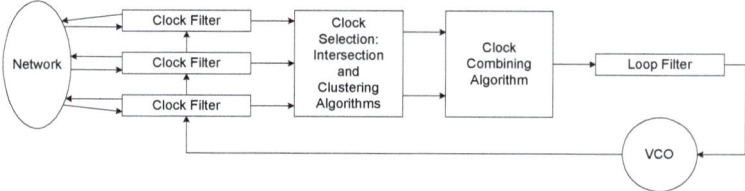

Figure 2.2: Network Time Protocol

meticulously described in [68]. It uses well-engineered statistical algorithms for data filtering and clock selection, employs a clock combining algorithm similar to [66] and consists primarily of a software phase locked loop that keeps the local clock in synchrony with an external time reference. The overall structure is illustrated in Fig. 2.2.

Clock Rate synchronization:
For tight clock synchronization in the *ns*-range oscillators with very low drift (e.g. OCXO's) are mandatory. The drawbacks of these devices are high cost, spacious design and high power consumption. To overcome these problems, a clock rate algorithm aims to reduce the drift term $P\rho$ without requiring an expensive oscillator at every node. Such an algorithm tries to adjust the clock rates in such a way that the rate differences can be bounded by a given consonance value γ, see Sec. 2.1. This value determines the amount a clock pair drifts apart during a re-synchronization period P, which is $P\gamma$ instead of the traditional $P\rho$. The goal of a rate algorithm hence is to achieve a γ that is much smaller than ρ.

In [95, 96] such a clock rate synchronization framework is presented. It relies on the same interval paradigm underlying the $O\mathcal{A}$ algorithm used for internal and external clock state correction. Amongst other things, the framework defines some generic convergence functions, that when applied reduce the influence of clock drift onto the overall achievable precision and accuracy. The major result is the following bound on the consonance:

$$\gamma = 6\sigma R + \frac{4\varepsilon}{R} \qquad (2.2)$$

Herein σ is the maximum oscillator stability, i.e. the maximum difference of the frequency measured at different times t_1 and t_2: $|f(t_2)/f(t_1) - 1| \leq \sigma$, see also Sec. 2.3.1. ε is the maximum uncertainty of the transmission delays and the parameter R is the re-synchronization period of the clock rate algorithm, which should be a multiple of the re-synchronization period P. It is important to point out that the worst case consonance γ of this rate algorithm does not depend on the oscillator drift ρ. To give a numerical example, let $\sigma = 0.0005\ \mu s/s^2$ for a TCXO, $\varepsilon = 400$ ns and $R = 30$ s, then $\gamma = 0.09\ \mu s/s + 0.053\ \mu s/s = 0.143\ \mu s/s$ by virtue of formula (2.2). This is a remarkable result in comparison to the drift of 1 $\mu s/s$ for clocks driven by a TCXO without any rate re-synchronization, see [96] for meticulous details.

2.3 Requirement analysis

In order to improve on the achievable precision and accuracy several parameters need to be addressed according to the previously given survey of clock synchronization algorithms. These parameters are for the oscillator, its drift ρ and stability σ, for the clock a state and rate correction mechanism, the re-synchronization period P, the clock granularity G, the clock setting granularity G_S and the clock rate adjustment uncertainty u. Of major interest is the clock reading error ε and the coupling to an external reference time. In the following subsections we will discuss and illustrate how these parameters are put together for typical system settings.

2.3.1 Clock Properties

In order to underpin the mechanisms involved, we briefly describe a typical clock design, provide typical oscillator parameters and sketch ways how clock corrections could be enforced.

The clocks in most computers nowadays are usually composed of an oscillator and an according Real-Time Clock (RTC) chip. The time-of-day is usually tracked in hardware within the RTC with a simple counter architecture that accumulates its registers holding the actual time with successive oscillator ticks. A battery backup is at hand to advance the time when the computer is switched-off, and the time can be set and adjusted via software. The simplified block diagram in Fig. 2.3 accounts for this principal clock architecture.

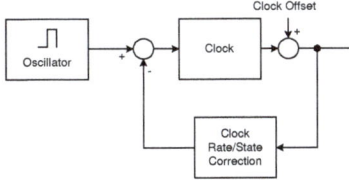

Figure 2.3: Computer clock control block diagram

Oscillator: In most computer systems the oscillator is an ordinary quartz crystal or crystal oscillator that drives a counter that can be offset via software and where the feedback block doesn't exist at all. Hence the oscillator indicates the progress of time with periodic ticks of a nominal frequency specified for the mounted device. The output of an oscillator can be expressed as

$$U(t) = (U_0 + \varepsilon(t))\sin(2\pi\nu_0 t + \phi(t))$$

where U_0 is the nominal peak output voltage, and ν_0 is the *nominal frequency* of the oscillator. The time variations of the amplitude are incorporated into $\varepsilon(t)$ and the time variations of the actual frequency, $\nu(t)$, are modelled by $\phi(t)$. The instantaneous frequency can be written as

$$\nu(t) = \nu_0 + \frac{1}{2\pi}\frac{d\phi(t)}{dt}.$$

For precision oscillators, the second term on the right-hand side is quite small, and it is useful to define the fractional frequency

$$y(t) = \frac{\nu(t) - \nu_0}{\nu_0} = \frac{1}{2\pi\nu_0}\frac{d\phi(t)}{dt} = \frac{dx(t)}{dt},$$

where

$$x(t) = \phi(t)/2\pi\nu_0$$

is the phase expressed in units of time. The difference between the instantaneous measure of the frequency from the nominal specified frequency for an oscillator is termed *oscillator accuracy*. Next to the accuracy, manufacturers specify their oscillators in terms of long term, short term and environmental frequency stability. More general, the frequency deviations of precision signal sources (i.e. oscillators) typically fall into two categories: systematic and random deviations. In non-precision oscillators, systematics typically dominate the frequency instability, whereas for precision oscillators random deviations are of major concern.

<u>Systematic deviations</u> are due to environmental effects (changes of the ambient temperature, supply voltage, load changes, pressure, ...) and can be expressed by

$$y(t) = y(t, T, U, Z, P, \ldots)$$

For small variations these effects can be separated and linearized resulting in

$$y(t) = y_0 + \mathcal{A}(t - t_0) + c_T(t - t_0) + c_U(t - t_0) + c_Z(t - t_0) + c_P(t - t_0) + \ldots$$

where $\mathcal{A}$ incorporates aging effects and $c_T, c_U, c_Z, c_P, \ldots$ are parameters accounting for different environmental effects. The following list surveys the effects that are usually specified in datasheets of COTS oscillators. These parameters are oftentimes specified following certain tests given in either the "MIL STD 202F" or the "IEC 60068" standards.

- Aging[3] is the systematic change in frequency with time due to internal changes in the oscillator and is hence a continuous measure of the oscillator accuracy, formally $\mathcal{A}(t) = \partial y(t)/\partial t$. At a constant temperature, aging has an approximately logarithmic dependence on time and is usually specified in $ppm/year$. The primary causes of crystal oscillator aging are stress relief in the mounting structure of the crystal unit, mass transfer to or from the resonator's surfaces due to adsorption or desorption of contamination, changes in the oscillator circuitry and slow changes of the crystal lattice. Generally, manufacturer specified long term stability includes oscillator aging but excludes environmentally induced effects.

- The static[4] frequency vs. temperature characteristics of crystal units are determined primarily by the angles of cut of the crystal plates with respect to the crystallographic axes of quartz, see [77]. Other factors that can effect this characteristic include the overtone, the geometry of the crystal plate, stresses of the electrodes and in the

---

[3]In contrast to aging, drift is the systematic change in frequency with time of an oscillator [43], meaning that drift is due to aging <u>and</u> changes in the environment and other factors external to the oscillator. Hence aging is what one denotes in a specification document and what one measures during oscillator evaluation, drift however is what one observes in an application.

[4]Static means that the rate of change of temperature is slow enough for the effects of temperature gradients to be negligible.

mounting structure, drive level, impurities and strains in the quartz material, ionizing radiation, the rate of change of temperature and thermal history [5]. Changing the temperature surrounding a crystal unit produces thermal gradients, which mainly cause the dynamic frequency vs. temperature characteristics. In general, the faster the temperature is changed, the larger is the contribution of the thermal-transient effect to the dynamic performance.

Almost every oscillator datasheet contains a specification of frequency stability over an operating temperature range in *ppm* or *ppb*, where the above mentioned factors are accumulated. For higher quality oscillators, to reduce these effects, the operating temperature is held constant above the ambient temperature with the help of an oven and a suitable control loop.

- Acceleration changes a crystal oscillator's frequency [109]. The acceleration can be a steady-state acceleration, vibration, shock, attitude change (2-g tipover), or acoustic noise. The amount of frequency change depends on the magnitude and direction of the acceleration $\vec{F}$, and on the acceleration sensitivity of the oscillator $\vec{\Gamma}$. The frequency change can be expressed as

$$\frac{\Delta f}{f} = \vec{\Gamma} \cdot \vec{F}.$$

Typical values of $|\Gamma|$ are in the range of $10^{-9}/g \ldots 10^{-10}/g$.

- Although quartz is diamagnetic, magnetic fields can change the frequency of an oscillator since the crystals mounting structure, electrodes and enclosure are affected. Time-varying electric fields will induce eddy currents in the metallic parts. Magnetic fields can also affect components such as inductors in the oscillator circuitry [9]. When a crystal oscillator is designed to minimize the effects of magnetic fields, the sensitivity can be much less than $10^{-10}/Oe$.

- Power-supply and load-impedance changes affect the oscillator circuitry and, indirectly, the crystal's drive level and load reactance. A change in load impedance changes the amplitude or phase of the signal reflected into the oscillator loop, which changes the phase (and frequency) of the oscillation [111]. The effects can be minimized through voltage regulation and the use of buffer amplifiers. Frequently found in datasheets are frequency stability for a 5% or 10% change of supply voltage and load respectively. The frequency of a "good" crystal oscillator changes less than $0.5 ppb$ for a 10% change in load impedance or supply voltage.

- Other influences on the systematic instabilities of an oscillator are due to ambient pressure, humidity, radiation, electric fields and gas permeation. Furthermore, the various influences on frequency stability can interact in ways that lead to erroneous test results if the interfering influence is not recognized during testing. For example, building vibrations can interfere with the measurement of short-term stability. Vibration levels of $\vec{F} = 10^{-2}g \ldots 10^{-3}g$ are commonly present in buildings. Therefore, if an oscillator's acceleration sensitivity is $\vec{\Gamma} = 1 * 10^{-9}/g$, then the building vibrations can contribute to the short-term instabilities at the $0.01 \ldots 0.001 ppb$ level.

Except for vibration, the short-term instabilities almost always result from noise. To characterize these *random deviations* the IEEE recommends either the measurement of the *spectral density of the phase fluctuations* $S_\phi(f)$ or the phase noise $L(f) \equiv S_\phi(f)/2$ in the frequency domain. In the time domain the *two-sample* or *Allan deviation*[5] $\sigma_y(\tau)$ is the measure of short-term instabilities, see [6].

A fundamental average value to characterize stochastic processes is given by the auto-correlation. For the steady-state this is obtained by

$$R_{yy}(\tau) = \int_{-\infty}^{+\infty} y(t)y(t+\tau)dt.$$

Using the relation from Wiener and Khintchine one can obtain the spectral density via fourier-transformation

$$R_{yy}(\tau) \circ\!\!-\!\!\bullet S_y(f)$$

where $S_y(f)$ is the spectral density of fractional frequency fluctuations in a 1-Hz bandwidth at fourier frequency f from the carrier v_0. With the relation

$$S_\phi(f) = \frac{v_0^2}{f^2} S_y(f) \quad [rad^2/Hz] \quad \forall \; 0 < f < \infty$$

one can obtain $S_\phi(f)$, the spectral density of phase fluctuations at frequency f from the carrier v_0.

In practice $S_\phi(f)$ is measured by passing $V(t)$ and $V_{ref}(t)$ through a phase detector and measuring the detector's output power spectrum:

$$S_\phi(f) = \left(\frac{V_{RMS}(f)}{V_S}\right)^2$$

where $V_{RMS}(f)$ is the root-mean-square noise voltage per $\sqrt{Hz}$ at Fourier frequency f, and V_S is the sensitivity in $[V/rad]$ at the phase quadrature output of the phase detector which compares the output of the "device-under-test" oscillator with those of a reference oscillator. In data-sheets $S_\phi(f)$ is usually denoted as phase noise at distinct frequency offsets given by:

$$S_\phi(f) = 20 log\left(\frac{V_{RMS}(f)}{V_S}\right) \quad [dBc/Hz].$$

Several measurement setups to directly derive these values are presented in [36] and [102].

In the time-domain the Allan deviation is defined by

$$\sigma_y(\tau) = \sqrt{\frac{1}{2}\langle(\bar{y}(t+\tau)-\bar{y}(t))^2\rangle}$$

where $\bar{y}(t+\tau)$ and $\bar{y}(t)$ are adjacent measurements of the fractional frequency deviation each averaged over a sample time τ. The expectation brackets "$\langle\rangle$" imply taking all possible values of over an infinite time average. For a finite set of N sequential adjacent samples

---

[5] The classical variance diverges for some commonly observed noise, such as random walk, i.e., the variance increases with increasing number of data points. In contrast, the Allan variance converges for all noise processes observed in precision oscillators, is easy to compute and faster and more accurate in estimating noise processes than the fourier transform.

of the frequency, each averaged over a sample time τ, one may estimate $\sigma_y(\tau)$

$$\sigma_y(\tau) \cong \sqrt{\frac{1}{2(N-1)} \sum_{k=1}^{N-1} (\bar{y}_{k+1}(t+\tau) - \bar{y}_k(t))^2}.$$

In order to quantify the systematic and random influences on the frequency stability of oscillators, Tab. 2.1 gives a comparison on salient oscillator parameters of some typical COTS oscillators extracted from various datasheets. A similar overview presented in Tab. 2.2 and taken from [110] specifies just ranges rather than actual product data.

	XO	VCXO	TCXO	MCXO	OCXO
Company	Quarz-Technik	AXTAL	Raltron	Temex	MTI
Type	TS-14/5	AXIS10	TX045	QEM77-AH	230-0666
Frequency stability vs. Temperature	±20ppm -20/+70C	±15ppm -20/+70C	±1ppm -20/+70C	±20ppb -30/+85C	±10ppb -30/+70C
Frequency stability vs. supply voltage change	±2ppm ±10%	±3ppm	±0.2ppm ±5%	±1ppb ±5%	±0.5ppb
Frequency stability vs. load change		±2ppm	±0.2ppm ±10%	±1ppb +1 gate	±0.5ppb
Stability/1 sec.				20ppb	0.1ppb
Stability/ 1 day				1ppb	0.5ppb
Stability/ 1 month				10ppb	
Stability/ 1 year	±5ppm	±3ppm	±1ppm		70ppb
Phase Noise [dBc]@1Hz					-95
[dBc]@10Hz		-80	-70		-125
[dBc]@100Hz		-110	-100		-145
[dBc]@1kHz		-135	-130		-150
[dBc]@10kHz		-145	-140		-160
[dBc]@100kHz			-140		-160
Size [cm^3]	1.36	0.82	2.1	19.3	19
Warmup Time		4ms			5min.
Power [W]	0.225	0.25	0.1	0.03	1.4/5

Table 2.1: Salient characteristics of COTS Quartz Oscillators (from datasheets)

	Quartz Oscillators			Atomic Oscillators		
	TCXO	MCXO	OCXO	Rubidium	RbXO	Cesium
Accuracy/year	2ppm	60ppb	10ppb	0.5ppb	0.7ppb	0.02ppb
Aging/year	0.5ppm	20ppb	5ppb	0.2ppb	0.2ppb	0
Frequency stability vs. Temperature	0.5ppm -55/+85C	30ppb -55/+85C	1ppb -55/+85C	0.3ppb -55/+68C	0.5ppb -55/+85C	0.02ppb -28/+65C
Stability/1 sec.	1ppb	0.3ppb	0.001ppb	0.003ppb	0.005ppb	0.05ppb
Size [cm^2]	10	50	20-200	800	1200	6000
Warmup Time [min]	0.1 (to 1ppm)	0.1 (to 20ppb)	4 (to 10ppb)	3 (to 0.5ppb)	3 (to 0.5ppb)	20 (to 0.02ppb)
Power [W]	0.05	0.04	0.6	20	0.65	30
Price (USD)	100	1000	2000	8000	10000	40000

Table 2.2: Comparison of frequency standards' salient characteristics (estimates)

From the data in Tab. 2.2 it can be seen that an improvement on the clock drift can be achieved by trading ordinary quartz crystal oscillators (XO's) with oscillators that are more stable and accurate, e.g. an ovenized quartz crystal oscillator (OCXO). Unfortunately, this also incurs some drawbacks, namely increased cost, power and space requirements and a long warmup time.

A different solution to cope with clock drift and clock instabilities is the utilization of a clock rate synchronization mechanism. The well engineered algorithm presented in [96] and [95] along with a rigorous analysis revealed that the clock rate stability (i.e. the maximum rate change per unit of time) takes over the role of maximum hardware drift rate in traditional clock synchronization approaches. With the values for the TCXO in the Tab. 2.2 this would mean a trade of the accuracy specified with $2ppm$ for the short term stability $1ppb$ that differ by a factor of more than 200. Hence when implementing a clock rate synchronization algorithm cheaper oscillators may satisfy the requirements for tight clock synchronization.

Clock: For clock synchronization some means to correct the local clock are required. Clock correction facilities can be either employed at the oscillator, the counting device or in software. A clock implemented in software on a host CPU is influenced by the operating system interaction and the CPU load, hence time should be maintained preferably with a hardware clock and means for correction should be made available in hardware as well.

Applying clock corrections at the oscillator requires a facility to vary the frequency of the oscillator itself:

- Some oscillators allow for a frequency adjustment via a potentiometer within a very narrow frequency range. With the help of a digital potentiometer the adjustments could be performed via a host CPU as well. Drawbacks of this method are the limited range for clock correction and the worsening influence of the potentiometer onto the clock drift and stability.

- More suitable devices, therefore, are *Microprocessor Compensated Crystal Oscillators* (MCXO), *Digitally Temperature Compensated Crystal Oscillators* (DTCXO) or *Voltage Controlled Crystal Oscillators* (VCXO). They offer a wider range on frequency adjustments and don't lack the worsening influence of the potentiometer. However, making an oscillator tunable over a wide frequency range degrades its stability because making an oscillator susceptible to intentional tuning also makes it susceptible to factors that result in unintentional tuning [110].

A different approach replaces the counter forming the actual clock by some hardware device that allows for clock corrections. Such a device provides greater flexibility and may support other features required for clock synchronization as well, e.g.:

- The pioneering *Clock Synchronization Unit* (AMI S65C60) omits or inserts single clock pulses at its clock register input, see [53]. This clock device requires an oscillator running at a multiple of the nominal clock frequency in order to allow for fine clock adjustments. Clock frequencies of about $100Mhz$ are required in order to achieve a smooth rate adjustment of $10^{-8}s/s$. Unfortunately, oscillators providing such high clock frequencies don't provide clock drifts and stabilities as those in the range of about $10-25MHz$.

- An adder-based clock forms the centerpiece of our *Universal Time Coordinated Clock Synchronization Unit* (UTCSU) clock chip. Employing an adder gives the freedom to add a particular amount (clock step) to the clock register at every pulse. A rate change can be achieved by varying this amount, which goes in effect almost instantly and holds up linearity. Figure 2.4 illustrates this simple technique for compensating a 10% slowdown of a clock. By extending the time register internally

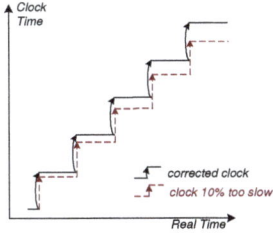

Figure 2.4: Adder based clock

with an ultra-fractional part it is possible to accumulate and correct time portions precisely. Apart from this adder-based clock principle our UTCSU Asic hosts a wealth of other features, see [97]. For implementation details the interested reader is referred to [61].

2.3.2 Clock Reading Error

Apart from clock drift, stability, granularity and rate adjustment uncertainty the clock reading error for both the local and remote clock is one of the main limiting factors for tight clock synchronization. Every node p in a distributed system that wishes to synchronize its clock to the clock of a remote node q and vice versa, needs to estimate the remote clock value. Two different ways to estimate remote clocks have been proposed so far:

  I. A *Clock Synchronization Packet* is sent periodically in a one-way fashion from node p to node q. Node q in turn estimates the remote clock value on node p based on timestamps that are piggy-backed onto this packet. This is possible when communication delays are bounded and clocks are initially synchronized, cf. Sec. 2.2.2.

  II. Remote clocks may be estimated with the help of round-trip packets as described earlier in Sec. 2.2.2 as well. This clock reading method provides clock values and error bounds and works in non-synchronous systems as well, however it doubles the number of messages.

The one-way time transfer usually delivers minimal uncertainty and is hence preferable for high accuracy clock synchronization. This method requires a mechanism to measure the constant and the variable transmission delays from node p to node q, which limit the accuracy for the estimation of a remote clock.

The following list details the steps involved in packet transmission/reception, see Fig. 2.5. Although these steps are derived from an Ethernet-based network, they apply to other networks as well (e.g. fieldbus systems). The contribution to the clock reading error of every step depends on the actual architecture and access policy of the chosen system. For the remainder we concentrate on Ethernet systems.

  (1.) The CPU at node p assembles a packet and stores it in an associated network buffer.

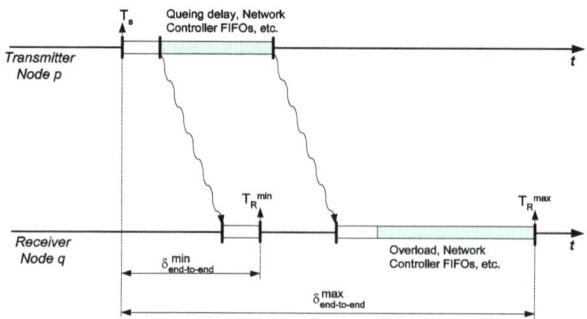

Figure 2.5: Remote Clock Reading Error

(2.) The CPU at node p signals its associated network controller to take over for transmission.

(3.) The network controller tries to acquire the network medium.

(4.) The network controller at node p reads the packet data from the buffer, serializes and encodes the data-stream and pushes the resulting bit-stream onto the medium.

(5.) The network interface at the receiving node q pulls the bit-stream from the medium, de-serializes the data and writes the packet into an according receive buffer structure.

(6.) The network controller at the receiving node q notifies its CPU of packet reception via interrupt.

(7.) The CPU at node q processes the packet.

Note that packet assembly and processing involves traversing of several layers of software since clock synchronization is usually built on top of some network protocol, e.g. the *Internet Protocol* (IP). An actual implementation of clock synchronization in software draws timestamps T_S and T_R respectively, when the clock synchronization algorithm is active — in the above list before the first and after the last item. For some scenarios the network controller at node p may find the network medium idle as well as the CPU at node q may actually process the receipt packet immediately, yielding a minimum end-to-end delay $\delta_{end-to-end}^{min}$. Other times a rather large end-to-end delay $\delta_{end-to-end}^{max}$ can be involved due to excessive queueing delays and CPU overload.

With bounded transmission delays the clock difference between the two nodes is given by

$$C_q(t) - C_p(t) \in T_R - T_S - [\delta - \varepsilon, \delta + \varepsilon]$$

with $\varepsilon = \delta_{end-to-end}^{max} - \delta_{end-to-end}^{min}$.

Providing a suitable timestamp mechanism which minimizes the transmission delay variability ε is crucial for tight clock synchronization. Hence timestamps should be triggered as close as practical to the physical layer. When properly implemented, the transmission delay variability ε can be reduced to the variability due to the physical layer connection ε_c and devices ε_d and due to synchronizer stages ε_s, which are required in asynchronous communications to bit-synchronize the data streams flowing from one physical clock domain to another. Usually, there are two clocking domains involved in serial transmission, a transmit and a receive clock domain, the latter being re-generated from the received serial datastream. A local clock device could use either one of these two clock domains (preferable) or an own clocking domain as its source. Thus, when timestamps should be drawn at a predefined edge of a bit cell within the clock synchronization packet, the different clocking domains need to be synchronized beforehand. For commonly used dual-stage flip-flop synchronizers $\varepsilon_s = 1/f$ with f denoting the frequency of the receiving clock domain, see [19]. To keep the error due to this sampling small, a sufficiently high clock frequency for over-sampling is mandatory.

Hence, for a network interface design one should try to minimize the amount of different clock domains and to maximize the clock frequency. The factors ε_c and ε_d however typically require experimental evaluation, due to the different physical layer devices and network topologies.

2.3.3 Clock Granularity an Clock Rate Adjustment

As illustrated in Fig. 2.3 the local clock of a node is assumed to be built upon a physical clock (usually driven by a quartz oscillator) of nonzero granularity G (micro-)seconds, which allows adjustment of rate and state. Therefore, clock ticks take place every (fixed) $G > 0$ logical time seconds. In practice, clock states can be adjusted at this logical times with a finite clock setting granularity G_S with $G_S \leq G$. Since in practice neither the oscillator frequency nor the clock resolution can be increased arbitrarily, every clock correction mechanism is bound to make some small errors due to G and G_S. However, most clock synchronization algorithms, except those developed in the SynUTC project [91, 87, 88], didn't consider these effects, since the remote clock reading error dominates by several magnitudes for COTS network interfaces without hardware timestamping capabilities.

Most clock designs correct local time at the clock rather than the oscillator, by tampering with raw oscillator ticks. These clocks tick at the intrinsic rate most of the time, whether they are adjusted or not. However, when the accumulated deviation between the intended and observed local time is about to exceed some bound u, the next tick is modified. This is enforced, e.g., either by pulse advancing/deletion, pulse suppresion/insertion or with an adder-based mechanism. This rate adjustment uncertainty u causes an additional uncertainty in the relation between logical time and real-time. The amount that a particular clock design contributes to G, G_S and u for different clock models were first presented in [91] and revised in [87]. For any kind of clock model the uncertainties due to the granularity and the rate adjustment uncertainty can be reduced by increasing the clock frequency f_{osc}. As an example, the UTCSU clock Asic provides a granularity $G_S = 444as$ that is internally extended to $1.73as$. In addition, when operating at $25Mhz$, it provides a rate error of $1.73as/40ns = 43.3 * 10^{-12}$, see [61]. Although these values would satisfy our needs, it provides timestamp registers for remote clock readings only with a granularity of $G = 2^{-24}s \approx 60ns$, which is clearly not sufficient for this kind of synchronization tightness.

2.3.4 Coupling to an External Reference Time

Tight coupling to an external time standard is required for external clock synchronization. In order to avoid a single point of failure coupling to different time sources that follow the official time standard UTC is desirable. A mechanism that is provided by most GPS receivers or receivers that follow other time sources as, e.g., LORAN-C or DCF-77, is the *one pulse-per-second* (*1 pps*) mechanism. Many GPS timing receivers provide an additional 10-MHz GPS output signal, that is derived from the frequency information received from the space vehicles atomic clocks. As long as GPS is up and running, this 10 MHz signal is very accurate[6] and best suited for sourcing a local hardware clock. In times of GPS outages, performance degrades to numbers inherent to the local GPS receiver oscillator design. Position, date and timestamps with a resolution of *1 s* are obtained through a serial interface while the exact start of each second is marked with either the rising or the falling edge of the *1 pps* signal. In addition, some GPS receiver designs use a dedicated signal which goes high or low if the receiver clock is locked to the satellite clock.

Summarizing the standard GPS timing interface that can be found on most COTS GPS receivers comprises a subset of the following hardware components and should be supported by any hardware support for clock synchronization:

- A bidirectional RS-232C port used to obtain position and time information and for configuration.

- A *1 pps* output or a Time-Mark output disciplined to GPS time or UTC.

- A 10 MHz reference frequency output disciplined to the satellites atomic clocks.

- A status line indicating the GPS receiver health condition (locked/unlocked).

2.4 Summary

This chapter established a general system model and identified key building blocks of existing clock synchronization algorithms. Next, relevant algorithms were informally described along with a listing of their achievable precision/accuracy. The addressed problem was to provide an overview of existing algorithms and to extract limiting parameters in order to identify possibilities for significant improvements, which is the primary aim of this thesis. The worst case analysis of the achievable synchronization precision π of the presented deterministic algorithms can be formulated by

$$\pi = c_1\varepsilon + c_2 P\rho + c_3 G + c_4 u + c_5 G_S \qquad (2.3)$$

where $c_1 \ldots c_5$ are small constants depending on the particular algorithm.

Herein,

1. ε denotes the transmission delay uncertainty dominating the remote clock reading error (*ms*-range for Ethernet using SW-based timestamping),

2. $P\rho$ denotes the clock drift during the resynchronization period (determined by the resynchronization period P and the oscillator drift ρ; $\mu s/s$-range for TCXOs),

3. G gives the clock granularity (resolution of clock readings), $u \leq G$ the rate adjustment uncertainty (timing error due to discrete rate adjustment; usually $u = 1/f_{osc}$) and

---

[6]A frequency stability of 10^{-11} over a 24 hours average and $5 \cdot 10^{-12}$ over a 7 day average can be achieved.

4. G_S accounts for the finite clock setting granularity.

Even tighter synchronizations can be achieved at the expense of determinism when using statistical or probabilistic algorithms.

When the clock drift $P\rho$ becomes the limiting factor better oscillators with smaller drift (OCXO's, etc.) become mandatory. Unfortunately, these devices have significant cost, size and power constraints that make them less suitable for commercial implementations. A clock rate correction can be used in this case to improve on the error due to the clock drift.

For providing accuracy information towards an external reference time adequate coupling of a suitable hardware clock to a suitable timing receiver is required. The last section of this chapter contained a requirement analysis in which these parameters, their significance and environment were described to some extent.

Chapter 3
Related Work

As stated in the introduction clock synchronization techniques may come in use in various ways. In pure hardware-based synchronization, dedicated clocking lines and clocking circuitry are employed to achieve a clock skew granularity down to the order of some *ns*. The extra interconnections are expensive and impractical for large systems and for systems with physically separated components. On the other end of the spectrum, software synchronization methods are generally used in loosely coupled systems where the order of achievable synchronization skew is in the range of some *ms*. Clock values are exchanged via message passing and used to synchronize the clocks at every node, hence no additional wiring and circuitry is required. If software synchronization alone is used, the latency and computational message overhead as well as message routing delays limit the achievable tightness of synchronization. System faults only exacerbate the message routing, traffic volume, and delivery time deviations. Unlike hardware techniques, the costs associated with software synchronization are related to the overhead of generating and handling the message traffic. Thus as a scalable general solution for distributed systems, synchronization based solely on either hardware or software techniques is often inefficient or impossible.

Alternatives to these conventional synchronization methods have been developed to enhance the fault resiliency and efficiency of synchronization, see [103]. Hybrid synchronization primitives are derived by combining software and hardware techniques and exploiting the benefits of each approach. A software model is typically superimposed over the system functions, with some functions off-loaded to dedicated hardware. This section briefly presents the clock synchronization concepts underlying different communication systems. Most bus-systems for safety critical embedded systems are time-triggered, employing a *time division multiple access* (TDMA) protocol. Hence, these systems specify an own, individual communications protocol, where clock synchronization is integrated and well established. Examples for these are the following bus systems employed and designed for avionic and automotive systems:

- *Multicomputer Architecture for Fault-Tolerance* (MAFT), a distributed system designed to provide reliable computation in launch vehicle avionics systems, see [48, 104].

- SAFEbus was developed by Honeywell ([37, 38]) to serve as the core of the Boeing 777 Airplane Information Management System, which supports several critical functions, such as cockpit displays and airplane data gateways. The bus has been standardized as ARINC 659 [41].

- Scalable Processor-Independent Design for Electromagnetic Resilience (SPIDER) is being developed at the NASA Langley Research Center as a research platform to explore recovery strategies for radiation-induced high-intensity radiated fields and/or electromagnetic interference (HIRF/EMI) faults, and to serve as a case study to exercise the recent design assurance guidelines for airborne electronic hardware (DO-254) [70].

- The Maintainable Real-Time System (MARS), a fault-tolerant distributed system for process control [49].

- The Time-Triggered Architecture (TTA), developed at the University of Technology Vienna, is deployed for safety-critical applications in cars and for flight-critical functions in aircraft and aircraft engines [50, 51].

- The FlexRay bus system is being developed by a consortium including BMW, DaimlerChrysler, Motorola, Philips and others. It is intended for powertrain and chassis control in cars [71].

Systems that solely propose concepts for clock synchronization superimposed onto existing, state-of-the-art communication protocols are:

- The NTI-module, an M-module for VME carrier CPU boards, developed in the course of the SynUTC project at the University of Technology Vienna [34] for Ethernet-based systems.

- Systems relying on the new IEEE 1588 standard [22].

- Extensions to several fieldbus systems, e.g., TTCAN, an extension of the Controller Area Network (CAN).

The following sections illustrate the most relevant concepts used for clock synchronization for some of these systems as they are publicly documented in scientific literature. Note that the concepts used in MAFT and SAFEbus rely on dedicated clocking lines, a mechanism that is not directly related to the concepts targeted in this thesis. SPIDER and FlexRay are still under development, hence only preliminary information is available. MARS, FlexRay and TTA do not use dedicated wires or signaling to communicate clock readings among the nodes attached to the network; instead they exploit the fact that communication is time triggered by a global schedule. Closest related to the concepts presented in this thesis are those used in the NTI-module design and the mechanisms proposed by the IEEE 1588 standard.

The next sections briefly present some of these systems and address how they handle the identified parameters, in particular the clock reading error and the coupling to an external reference time, as outlined in the previous chapter.

3.1 MARS - The Maintainable Real-Time System

An engineered hybrid clock synchronization approach is built into the distributed fault-tolerant real-time system MARS [49]. The system architecture strictly separates issues of synchronization and timeliness, data transformation, and the dependability aspects (e.g., error detection, error handling and redundancy management). This is enforced by implementing and instantiating all system activities in a *time-triggered* fashion. The MARS system is built on the concept of clustering, where each cluster consists of several nodes interconnected by a synchronous real-time bus run by a *time division multiple access* (TDMA) protocol. TDMA guarantees each component to have access to the medium at equidistant points in time, called *TDMA slots*. Hence, there is a constant number of messages every node may send in a time interval. Dedicated TDMA slots are used for exchanging clock synchronization messages. Here every message contains the timestamp of the sender's clock and the receiving node attaches the timestamp of the receiver's clock to the incoming message. Every node records the time differences to the other nodes periodically. Based on this information, a correction term for the local clock is calculated with the Fault-Tolerant Average Algorithm that is applied to the local clock.

The clock synchronization concept of MARS first introduced the following significant innovations:

- Memory-mapped timestamping is used to reduce the remote clock reading error.

- The Clock Synchronization Unit (CSU), maintains the local clock and provides facilities for clock correction, cf. [73, 53]

- External clock synchronization is achieved by coupling one node to an external reference timing receiver.

Figure 3.1 schematically shows the timestamp mechanism for messages in MARS. The CPU at the sending node p places a waiting message in a transmit buffer within its memory and signals the network controller to start with transmission. The network controller in turn fetches the packet from the memory by direct memory access (DMA) and streams the serialized data onto the MARS bus along with an associated TDMA slot. This method exploits the capability of the network controller to package several memory fragments continuously into one message for timestamping. In particular the last fragment of each message is a memory-mapped, real-time register of the CSU clock chip that is accessed at the moment of sending. At the receiving node q an interrupt is issued by the network controller immediately after a message arrives. This interrupt is directed to the CSU clock chip at the receiver, which generates a timestamp. Afterwards, the CPU reads this timestamp from the CSU register and copies it to the according receiver message buffer.

Varying delays due to the network controller internal FIFO structures and arbitration latencies at the receiving node dominate the contribution to the remote clock reading error. Note that the arbitration delay can become quite large when the network controller is not able to arbitrate the memory at the receiver, e.g., if the CPU uses the memory. Furthermore, it must be ensured that the received timestamp is read off the CSU before a second one is drawn.

The CSU maintains local ($G \sim 1\mu s$) and global ($G \sim 100\mu s$) time with a counter-based approach. The clock times are continuously rate and state correctable with the help of correction terms that can be set in the range from $100ns$ to $10ms$. Clock correction is enforced by insertion/suppression of clock ticks. Therefore, the CSU must operate from a multiple of the nominal clock frequency in order to allow for fine adjustments. The drawback of this technique is not only a higher required bandwidth, but also an unsatisfactory correction granularity (a small rate change means a long delay until one pulse gets modified).

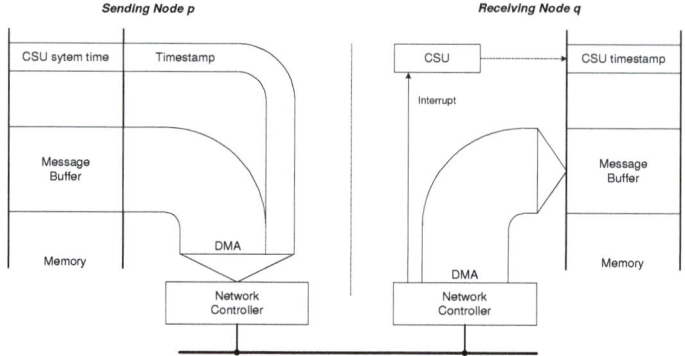

Figure 3.1: Timestamp mechanism in MARS

Furthermore, it contains a sample register for accurate determination of an interrupt signal that is dedicated to sample the arrival time of an incoming message. A DMA-like host interface is used to draw and transparently map a timestamp into an outgoing message after the media access to the network has been granted.

To facilitate external synchronization every MARS cluster contains a node with access to an external time standard that measures the deviation between the cluster's time and those of the standard. An external clock synchronization task broadcasts an appropriate rate correction that affects the speed of all internal clocks independently of corrections due to internal clock synchronization. However, the CSU has no interface for direct coupling of an external timing receiver, hence host CPU interaction is required, a fact that worsens the achievable accuracy.

In [49] the authors state that the MARS implementation, where the TDMA protocol is run atop a 10 Mbps shared media Ethernet interface, achieves a synchronization tightness below 10μs.

3.2 The Time-Triggered Protocol

The Time-Triggered Protocol (TTP) was designed to meet the stringent requirements for distributed fault-tolerant control systems, see [50]. Currently there exist two available implementations of TTP:

- The TTP/A protocol is a low-cost member of the TTP protocol family that is intended for non fault-tolerant field-bus applications. It's primary usage is to connect a node with sensors and actuators.

- The TTP/C protocol is the full version of the time triggered protocol, that provides several services required by a real-time bus of a fault-tolerant distributed system.

Every node consists of a host processor performing the application specific task and a communication controller executing the TTP/C protocol. These two parts are interconnected by the *communication network interface* (CNI) implemented with the help of a dual-ported memory. Additionally, every node contains a bus guardian. The bus guardian ensures the correct behavior of a node in the time domain by controlling the bus access of the TTP/C communication controller. Upon a violation of the access pattern it terminates the controller operation in order to fulfil the fail-silent condition.

Several nodes can be interconnected with two redundant communication channels employing different bus structures (star, linear bus, etc.). A set of these nodes forms a cluster, where every node can communicate directly by way of broadcast messages using the time triggered communication protocol TTP/C. Clusters may be connected via gateway nodes to allow communication between nodes of different clusters, that is inter-cluster communication. In addition, access to an external reference time can be realized via time gateway nodes.

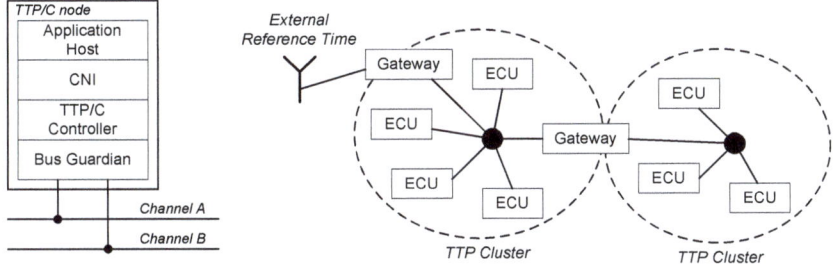

Figure 3.2: TTP/C Node (left) and System (right) Architecture

The medium access strategy of TTP is based on a static TDMA scheme. The total channel capacity is divided into a number of slots that are statically assigned to the nodes. This a-priori knowledge is used for the efficient implementation of several protocol services.

One particular service is a fault-tolerant global time base of known precision at all nodes. Therefore, TTP provides a fault tolerant internal synchronization of the local clocks without any overhead in frame length. Due to the static TDMA cycle receivers know a-priori the point in time each message is transmitted. Assuming the transmission latency is constant the deviation of the pre-specified send time and the observed receive time indicates the difference between the sender's and receiver's clock. Hence, it is not necessary to exchange explicit synchronization messages or to carry the value of the send time in the message which would extend the message length. Thus every node contains a local oscillator that ticks with a frequency determined by its physical parameters. A subsequent number of these node-local microticks form the so-called macrotick which is used to increment the global time counter of the node. In order to establish a global timebase with the specified precision the macroticks of each node must be re-synchronized periodically with the operational ensemble. The macrotick generation of the local node (=global timebase) can be influenced by changing the number of microticks per macrotick. This timebase is maintained by an integrated hardware clock design, embedded in the TTP/C controller, that follows the principles and mechanisms derived in the MARS project, see [52]. Clock synchronization is achieved by measuring the difference between the expected arrival time of a message and its actual arrival time, determined by the signal edge of the start of a new message. Note that the expected arrival time is a function of the receiver clock and the actual arrival time is in direct relation with the sender clock. The difference of both seen at the receiver is measured with the granularity of the receiver clock that directly relates to the achievable clock precision. A *fault-tolerant average* algorithm is employed to correct the clocks at every node. The achievable precision here is given by

$$\pi = (\varepsilon + 2\rho R)((N-2k)/(N-3k))$$

for N clocks within the system where k clocks may exhibit Byzantine faults. Again the major contribution is due to the clock reading error that is determined by the variability in the edge detection of the start of a new message plus the variability in the message transmission times. Therefore, the rates of both clocks as well as the transmission speed of the link dominate this value. For an IP-module prototype implementation running at a bitrate of 100 kb/s a precision of $2.356\mu s$ and a granularity of $3.8\mu s$ were measured, see [52].

External clock synchronization is not part of the TTP/C protocol but can be added by giving a node access to an external time base. A *time gateway*, see Fig. 3.2, forces its view of external time on all its subordinates by periodically sending a broadcast time message. However, in order to avoid relative time measurement errors due to a malicious time gateway node the time server may only have a limited authority to correct the clock rate of a cluster.

The reading error ε is determined by the variability in the edge detection of the start of a new message plus the variability in the message transmission times. This variability is determined by the physical shape of the signal on the transmission channel and the granularity of the local time measurement (the microticks) in the node that records the edge detection. Minor effects, e.g., the clock rate adjustment uncertainty as identified in [85] are of no/small concern in TTP systems, since the achievable precision is targeted to the μs-range. More troublesome becomes the clock drift that is related to the precision with a factor of $2\rho R$. Since the re-synchronization period R depends on the static TDMA layout, this factor will degrade π, when a cluster is composed of many nodes, hence this should be kept in mind during system design. Furthermore, the support for coupling external reference timing receivers has several deficiencies:

- The TTP/C controller provides neither a *1pps* input nor a corresponding serial interface for timing receivers.

- Processing of this timing information is usually a task of the application host.

- Coupling of multiple reference timing receivers is not directly supported in hard+firmware either.

Concerning clock synchronization the FlexRay system is very similar to the TTP system, therefore the addressed items in general apply to this system as well. Differences arise due to the different bus protocol that is a mixture of static TDMA and dynamic concurrent access mechanisms, see [71].

3.3 The Network Time Interface

The *Network Time Interface* (NTI) was developed in the SynUTC project, see [33, 34, 89]. It was designed to support fault tolerant external clock synchronization in distributed systems that are based on standard, packet-oriented networks.

In the SynUTC approach, every node has to be equipped with a hardware clock, in our case the *Universal Time Coordinated Synchronization Unit* (UTCSU) [97], a general purpose CPU responsible for executing the software part of the clock synchronization algorithm, and a Communication Coprocessor, which provides access to the network by reading/writing data packets from/to (shared) memory independently of CPU operation (e.g. via DMA). For external synchronization purposes, some nodes need to be provided with external time sources like GPS satellite receivers. Although such nodes have additional functionalities, their hardware architecture remains the same.

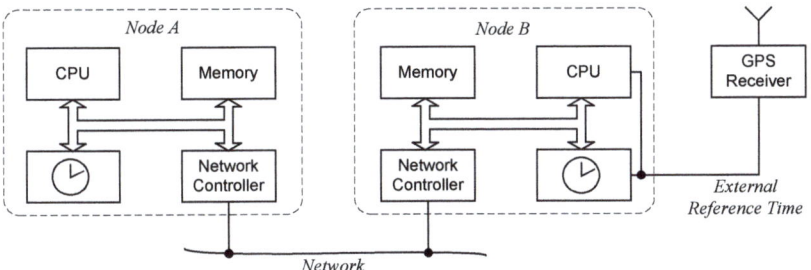

Figure 3.3: SynUTC: System Structure and NTI Architecture

The NTI built as an M-Module contains the clock (UTCSU), dedicated memory for packet transmission/reception and some glue logic required for transparent timestamping. The mechanism for the exchange of clock values is a refinement of the principle used in the MARS project. In particular, it provides timestamp facilities whenever the network controller grabs a packet for transmission from the memory and when a packet is received and written to the memory respectively. Here a dedicated address decoding logic was used to trigger timestamps and to map the transmit timestamp transparently into the outgoing datastream whenever the network controller reads a predefined address location within the local memory dedicated for this purpose. On packet reception the receive timestamp is copied in the receive interrupt service routine from the according UTCSU register to the dedicated memory location within the packet. For both situations the required fields within the packet payload are overwritten by this particular mechanism, hence the according clock synchronization driver software has to account for this circumstance. In contrast to the approach taken in the MARS project the timestamps are here inserted at the start of the payload, see Fig. 3.4, and the trigger position in time of the timestamps can be adjusted to allow for an optimization of the implementation. Furthermore, this method doesn't need to misuse the receive interrupt request for triggering of receive timestamps, since the interrupt is usually required for several other purposes as well and would hence trigger additional timestamps rather frequently.

The NTI profits of the wealth of functionality of the custom *Universal Time Coordinated Synchronization Unit* (UTCSU), see [97, 61]. Due to its flexible bus interface,

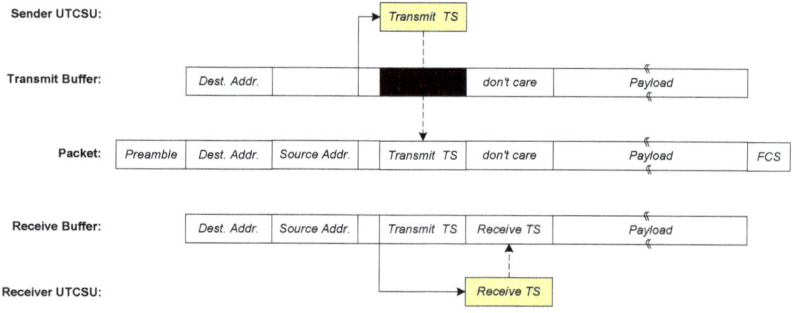

Figure 3.4: NTI Timestamping

featuring dynamic bus sizing and little/big-endian byte ordering, the UTCSU can be used in conjunction with virtually any 8, 16 and 32 bit CPU. Figure 3 gives an overview of the major functional blocks inside the UTCSU. Fig. 3.5 shows a block-diagram of the major functional blocks inside the UTCSU.

The centerpiece of the UTCSU is the local clock implemented as a 91 bit wide adder within the Local Time Unit. The most significant part utilizes a 56 bit NTP-time format which maintains a fixed point representation of the current time with a 32 bit integer part and a 24 bit fractional part. Clock time can be read atomically with a resolution of $2^{-24} \approx 60ns$. The local clock of the UTCSU can be driven by any oscillator frequency f_{osc} in the range of $1\ldots 25MHz$, is fine-grained rate adjustable in steps of about $10ns/s$, and supports state adjustment via continuous amortization.

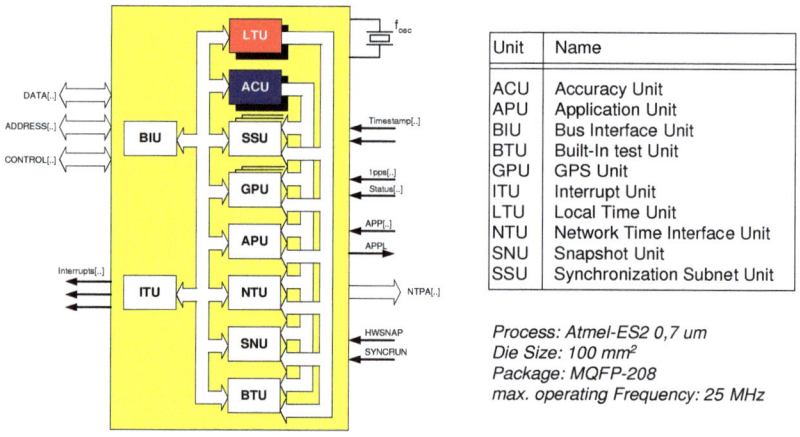

Figure 3.5: UTCSU block-diagram and technical data

To achieve both internal and external clock synchronization, the NTI approach relies on an interval-based paradigm: Real-time t (usually UTC) is not just represented by a single time-dependent clock value $C(t)$ here, but rather by an accuracy interval $A(t)$ that must satisfy $t \in A(t)$. More specifically, accuracy intervals are provided by combining an ordinary clock $C(t)$ with a timedependent interval of accuracies $[-\alpha^-(t), \alpha^+(t)]$ taken relatively to the clock's value, leading to $A(t) = [C(t) - \alpha^-(t), C(t) + \alpha^+(t)]$. Since accuracy intervals need to be maintained dynamically, they are quite small on average. For an in-depth treatment of system modelling and clock synchronization algorithms based on this paradigm consult [91].

In order to support interval-based clock synchronization, the UTCSU contains two more adder-based "clocks" in the Accuracy Units that are also driven by the oscillator frequency f_{osc}. They are responsible for holding and automatically deteriorating the 16 bit accuracies $\alpha^-(t)$ and $\alpha^+(t)$ to account for the maximum oscillator drift. Both can be (re)initialized atomically in conjunction with the clock register in the LTU. In addition, some extra logic suppresses a wrap-around and zero-masks potentially negative accuracies during continuous amortization.

Several trigger signals sample the current local time/accuracy into dedicated UTCSU registers in the Synchronization-Subnet Units, GPS Units and an the Application Unit. The manyfold implementation of these units facilitates redundant communication architectures and/or gateway nodes. Two of these trigger signals serve on the NTI as transmit and receive timestamp triggers respectively. Furthermore, several other trigger signals allow direct coupling of GPS receivers via a dedicated *one pulse per second* (1pps) and an optional *status* signal.

A thorough system evaluation presented in [90] and [89] gives a worst case accuracy/precision in the $10\mu s$ range (average case $1\mu s$). In addition an analysis on the measured data, in particular the clock reading error ε, revealed several factors that limit tighter synchronization:

- Arbitration latencies and transmission speed limitations of the involved bus interfaces between the CPU, the memory and the network controller.

- Uncertainties caused due to network controller internal peculiarities.

3.4 IEEE Standard 1588

IEEE 1588 is a new standard –approved in 2002– for a precision clock synchronization protocol for networked measurement and control systems. The specified *precision time protocol* (PTP) is applicable to systems communicating by local area networks supporting multicast messaging within subnets including, but not limited to, Ethernet. Hence, in a PTP system the nodes participating in clock synchronization can be interconnected in various different ways, e.g., direct or via repeaters, switches and/or routers. Provisions are made to minimize delay variations incurred in the message transmission used for clock synchronization between any sending and receiving timestamp. This is enforced by the requirement to generate timestamps as close to the physical layer as is practical for a given clock implementation. Small delay fluctuations introduced by the PTP protocol stack and by network components (e.g., due to repeaters in an end-to-end communication path) can be reduced by averaging. When switches or routers are in-between an end-to-end path, these variations may become too large. Therefore, PTP specifies a boundary clock mechanism that passes-by these elements. In addition, boundary clocks are mandatory when synchronizing across subnets.

PTP specifies a master-slave structure using a best-master-clock selection mechanism to dynamically select the master within a subnet. All slaves subsequently synchronize their clocks to those of the selected master. A master clock selected across several subnets is termed grand-master. Exchange and selection of clock values involves the following steps:

1. A node in the master state sends a multicast SYNC packet to all slaves containing an estimate of the sending time and characterization information of the master clock.

2. The timestamp logic at the master node senses this outbound packet close to the physical layer and triggers a precise send timestamp from the master nodes *Real-Time-Clock* (RTC).

3. The timestamp logic at every slave node senses the inbound SYNC packet and triggers a timestamp from the slaves *Real-Time-Clock* (RTC).

4. The master sends a follow-up packet associated with the preceding SYNC message containing the previously precise send timestamp to all slaves.

These steps are also performed in the reverse direction, although less frequently, using Delay_Req and Delay_Resp messages. Using this information the one-way transfer delay can be estimated. The slaves succinctly use this information to correct their clock to those of the master. In [22] a simple PI-controller is used as clock correction mechanism. Simple experiments based on a Fast Ethernet implementation, in different configurations, e.g. direct link using a cross-over cable and connection of two nodes via a repeater and a switch, respectively, were conducted. These experiments, where both a simple oscillator and an expensive OCXO were used, yielded standard deviations between the master and the slave clock in the range from 30 to 150 ns depending on the actual configuration under low network-load conditions. In particular, when repeaters or switches are used under loaded conditions, the fluctuations will get worse. For routers this is true even for an unloaded case. To overcome these problems PTP suggests[1] the implementation of boundary clocks to serve as a time transfer mechanism between subnets. Therefore, the router must be configured to block all IEEE 1588 messages. The boundary clock has a network connection to each of the subnets. When viewed from a subnet the boundary clock appears exactly like any other (ordinary) 1588 clock in the system. Within a subnet the ordinary clocks and the portion of the boundary clock visible from the subnet synchronize with each other as though they were all ordinary clocks. The boundary clock itself resolves all of the times of the several subnets by establishing a parent-child hierarchy of clocks. In a system with a single IEEE 1588 Boundary Clock, the boundary clock will typically be at the root of this hierarchy and will be the master clock for all of the clocks in each of the subnets. In addition to the synchronization functionality, an IEEE 1588 Boundary Clock should provide a retransmission mechanism for 1588 management messages.

If one analyzes IEEE 1588 the following pros and cons can be identified amongst others:

+ The provision of timestamping as close to the physical layer as practical.

---
[1]Currently, the standard doesn't specify boundary clock support for switches.

- PTP has the potential to become a robust, self-configuring protocol that possibly can support a multitude of different bus systems. (Its applicability for the specified bus systems —fieldbusses, PC-based bus-systems, etc.— need to be analyzed and elaborated in more detail.)

- The boundary clock mechanism can minimize delay fluctuations introduced by routers. This mechanism can possibly be extended to switches as well in order to achieve a tighter accuracy.

- The time representation and the timestamps with 64 bits are not sufficient to cover granularity and rate adjustment effects as outlined in [85].

- The master-slave approach outlined in PTP has several shortcomings concerning fault-tolerance and limits the applicable synchronization algorithms.

- The delay between a SYNC message and a corresponding follow-up packet can degrade the achievable precision due to freewheeling clocks at either side. To that end, the error due to the drift should be corrected.

3.5 Summary

This chapter presented the hardware support and principles of clock synchronization that are involved as specified in different distributed systems. The four described systems use hardware timestamping to tackle the clock reading error. In the MARS and the SynUTC-NTI implementations timestamps are transparently inserted into in- and outbound packets at the memory interface between host and the network controller, whenever the network controller accesses the packet. IEEE 1588 in contrast samples timestamps as close as practical to the physical layer, timestamps of outbound packets are sent with a follow-up packet to the receiver. TTP and FlexRay also timestamp next to the physical layer. But in contrast to 1588, they do not exchange timestamps for internal synchronization at all, instead they use deviations from the global TDMA schedule to compute clock differences to other nodes. This mechanism is especially useful for cost sensitive applications and when a synchronization accuracy above $1\mu s$ is sufficient.

The main focus of hardware support for clock synchronization is the implementation of the clock itself. In the MARS project the clock synchronization unit (CSU) maintains the local clock with a counter. Clock corrections are provided via a mechanism for insertion/suppression of clock pulses. An advanced approach was implemented by the SynUTC UTCSU clock Asic. Here an adder-based clock is used that allows for smooth corrections by adding/subtracting fractions of a second. In addition, the UTCSU provides support for external clock synchronization by direct coupling to timing receivers. The external reference time is tracked with the help of accuracy intervals that specify a maximum deviation between the local and the reference time.

Chapter 4
Network interface architectures supporting tight clock synchronization

In the previous chapters we derived several parameters that need to be addressed when hardware support for high accuracy clock synchronization over packet oriented networks is implemented. The following aspects that need to be considered in order to achieve a synchronization precision/accuracy in the range of some *ns*:

State adjustments: The clock at each node should be able to perform state adjustments for ironing out non-systematic, short-term clock deviations that arise after power-up, node join and during system operation. Every node broadcasts periodically its current clock state and determines from all incoming clock states a new one with enhanced quality. To avoid non-monotonic clock correction, continuous amortization periods should be used for adjusting the local clock state accordingly. More specifically, a node performs the following operations:

- **Initiation:** In a periodic fashion the nodes participating in clock synchronization initiate a broadcast of their current clock state.

- **Sending:** Due to computational overhead and contention in the communication medium, a node is able to start sending the clock state only after some delay. Every clock synchronization packet gets timestamped on actual departure.

- **Collection:** A node receives clock synchronization packets from all other nodes in a sequential manner, timestamps their arrival and stores them in a suitable data structure.

- **Updating:** Since clock synchronization packets are sent and received at different points in time, the collected clock states have to be made compatible with each other.

- **Termination:** Some fixed time after initiation chosen appropriately so that all clock synchronization packets from other correct nodes have been received, final re-synchronization activities are started.

- **Computation:** Based on the remote and the local clock states a new correct clock state is computed by an appropriate convergence function, see Sec. 2.2.2 for some examples.

- **Amortization:** The local clock gets adjusted in a smooth way according to the correction value.

Rate adjustments: Since clock drift influences the achievable precision by at least $2\rho R$, where ρ denotes the oscillator drift and R the re-synchronization period, some mechanism is required to limit this influence. Employing an oscillator with small drift could be a possible solution here. Unfortunately, OCXO's or rubidium oscillators are very expensive and consume much power and space. An alternative solution is the implementation of a clock rate algorithm, see [96], that allows to trade the clock drift by its stability, which provides an improvement of a factor of $10^2 \ldots 10^3$, see Tab. 2.2. The operations required for a clock rate algorithm are similar to those used for a clock state algorithm, hence an implementation is very cheap in terms of additional communication as well as computation cost.

Clock granularity: Detailed analysis of the *orthogonal precision* and the *orthogonal accuracy* algorithm, see [88] and [87], revealed that clock granularity G and rate adjustment uncertainty u influence the achievable worst-case precision/accuracy. Advanced clock circuitry, like the UTCSU, provide a variable internal granularity and utilize an "artificial" rate, generated by discrete rate adjustment techniques. For a precision/accuracy in the *ns*-range these properties shall be devised properly.

Message timestamping : The lower bounds given in [23] clearly state that the clock reading error is of utmost importance when deterministic clock synchronization is implemented. Experiences made with the implementation of our NTI module led to the conclusion that timestamp facilities required for remote clock reading should be ideally placed next to the physical layer of a network interface in order to avoid transmission delay uncertainties introduced due to network controller internal circuitry, see [89]. Unfortunately, placing the clock circuitry between the network access layer and an according physical layer degrades the efficiency in application-level clock reading/programming, hence considerable thought should be spent on "architecting" this interface.

Coupling of external time sources: Fault tolerant synchronization to external time sources is required to facilitate external clock synchronization. Potential receiver errors and coupling errors directly relate to the achievable accuracy, see [29]. Hence, a dedicated mechanism to interface to receivers that provide access to an external time standard, e.g., GPS receivers, is required. A similar approach as taken by the UTCSU to timestamp the *one-pulse-per-second* signal seems appropriate here, too, although the timestamps should be extended to match the required granularity.

In order to mask potential receiver failures or rare faults caused by the time sources themselves, redundant coupling to different external time sources is mandatory. This can be done by coupling these receivers to different dedicated nodes within the system, thus allowing to mask also potential geographical restrictions due to surrounding obstacles.

Bus interfaces: As revealed in [89], bus interfaces can become an often unexpected bottleneck and may impair significantly the achievable precision/accuracy. Apart from throughput aspects the interfaces should comply with a standard to restrict an implementation to specialized technology.

Putting all these requirements together the following sections present our patent pending architecture [47] that is tailored for Ethernet technology.

4.1 System Architecture

When support for clock synchronization is added to an existing network, the added hardware and software should keep modifications to the actual implementations as small as possible. In particular, to allow for a smooth migration from an un-synchronized to a synchronized network it is mandatory that all functions should operate the same way prior and afterwards.

A careful analysis of all relevant delays and delay variation that incur in an end-to-end path should be undertaken since these parameters limit the achievable precision/accuracy.

Fig. 4.1 illustrates a typical office network that forms a part of the lowest hierarchy of the data network at the University of Technology Vienna.

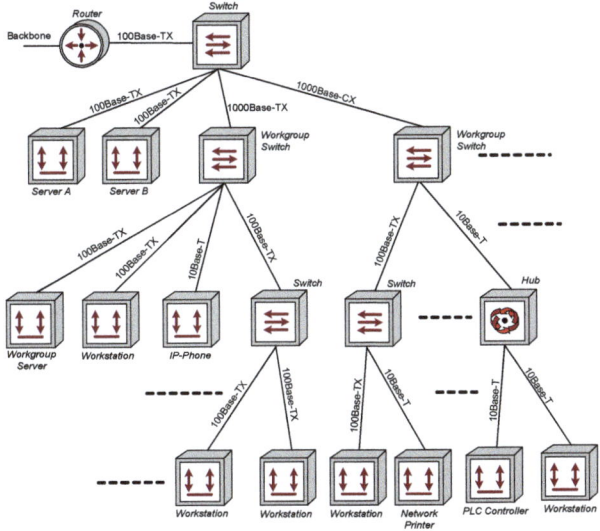

Figure 4.1: A typical office network topology

As can be deduced from Fig. 4.1 such a network consists of different cabling technologies (twisted pair, fibre-optic, etc.), end-systems and several devices to interconnect these end-systems (hubs, switches, routers, etc.). Different cabling types and lengths as well as different physical interfaces introduce different delays in every segment.

Data Terminal Equipment is any source or sink of data connected to the local area network. Typically, these are the computers that form the end-systems in an end-to-end communications link. Delay and according fluctuations are due to the hard- and software. Tab. 4.1 presents an expected order of magnitude of the expected delay variations extracted from an experimental evaluation presented in [45] for COTS PC's running Linux.

Ethernet cable: Within the different Ethernet varieties, different cable types are in use. For copper media the delay is roughly proportional to the square root of the dielectric

Domain	Expected Average Delay Variation
SW: Application, Operating System, Protocol Stack, Device Driver	$\geq \mu s$-range
HW: Medium Access Controller, Physical Layer Interface	$\geq 10ns$-range $\geq ns$-range

Table 4.1: Average delay variations of a DTE (from [45])

parameter $\sim c_0/\sqrt{\varepsilon_r}$. Tab. 4.2 gives a short overview of the delay variations that are encountered by the different Ethernet cable types:

Cable Type	Delay	Delay Variation
Twisted Pair CAT-5 UTP	max. allowed 5.7ns/m typical 5ns/m	$< 1ns$
1Base-5, 10Base-T, 100Base-T and 1000Base-T		
Optical fiber	max. allowed $5.05ns/m$	$< 1ns$
110Base-F, 100Base-FX, 1000Base-CX and 1000Base-SX		
Coaxial cable 10Base-2 10Base-5	max. allowed $6.3ns/m$ max. allowed $7.5ns/m$	$< 8ns$
10Base-2, 10Base-5 and 10Broad-36		

Table 4.2: Fluctuations due to Ethernet cable types (see [39])

Ethernet hubs, also termed multiport repeaters, operate at the Physical Layer of the OSI Reference model. They are used to connect one or more Ethernet cable segments of any media type. If an Ethernet segment were allowed to exceed the maximum length or the maximum number of attached systems to the segment, the signal quality would deteriorate. Hubs and repeaters are used between a pair of segments to provide signal amplification, timing and preamble regeneration to restore a good signal level before forwarding the frames.

According to their implementation, they add certain delay and delay variation (*jitter*) to the transmission of packets. Tab. 4.3 gives the maximum allowable delays and delay variations for 100Mb/s baseband networks as specified in [39]. The values are given in *bit times* (BT). A bit time is the duration of one bit as transferred to/from the MAC and is the reciprocal of the bit rate. The bit time for 100Base networks is $10^{-8}s$ or $10ns$. The values

	Class I repeater	Class II repeater	Variability
100Base-FX	$\leq 140BT$	$\leq 46BT$	$7BT$
100Base-TX	$\leq 140BT$	$\leq 46BT$	$7BT$
100Base-T4		$\leq 67BT$	$8BT$
100Base-T2		$\leq 90BT$	$8BT$

Table 4.3: Maximum allowable Repeater delays for 100 Mb/s

given in Tab. 4.3 show that every hub can add a worst-case delay variation of $80ns$ onto the packet transmission delay.

Ethernet switch, also termed multiport bridge, is a LAN interconnection device. Newest brand switches operate at layers 4 of the OSI model; while the devices that are most often employed nowadays operate at layer 2 or 3. Switches process all packets received on any port and forward them to an associated outgoing port. Hence, a switch makes it possible to filter traffic passing between its ports.

In spite of its capability to switch port connections, there can be significant delay in forwarding frames to output ports when they are congested. In fact packets are either stored or dropped when the required buffers become insufficient. Since the principal function of a switch is to bridge frames between its ports, it tries to do this as quickly as possible, in terms of throughput by achieving the maximum bandwidth but also in terms of delay. A switch makes its time-critical filter and forward decision based on the destination address and/or the frame integrity. In general, two major categories can be distinguished:

- **Store-and-forward** allows for full error checking, packet filtering and LAN speed conversions at the cost of higher transit delay, especially for large packets. An entire packet needs to be received and processed before it is forwarded.

- **Cut-through** minimizes transit delay by foregoing the possibility of error checking, packet filtering and speed conversion. Packets are forwarded as soon as the destination address has been received.

- **Fragment-free mode** is cut-through switching in which runt packets (collision by-products of less than the minimum legal packet size of 64 bytes) are discarded.

There has been no typical trend to one distinct method, on the contrary, today most switches provide multiple, selectable modes [98].

The design of the internal switch fabric is critical to the performance of a switch. There are two mainstream switch fabric architectures that have been widely used in commercial LAN switch products:

- **Shared memory** architectures are very common for low cost, small scale switches and have the advantage of easily accommodating mixed LAN types and speeds within a single switch. Shared media switches use a high-speed backplane to interconnect switching elements, which may consist of an individual bridge per port or a multiport switch module. The latter may use shared media or shared memory as an internal architecture. Shared media architectures are frequently used to build modular switches that can scale to high port densities.

- **Cross-point matrix** switches employ an array of switching elements to provide parallel switched paths between distinct pairs of input and output ports. This design approach has yielded a fairly attractive price per port in Ethernet switches with relatively few ports.

Each approach reflects a distinct method used to move frames from input to output ports and has its own characteristics, limitations, and design issues. A comprehensive treatment of switches and more details on their architecture can be found in [98].

Fig. 4.2 illustrates the different transit delay principles that are encountered by store-and-forward and cut-through techniques. Typical values for the transit delay are $\sim 40\mu s$ for cut-through and $\sim 10 - 150\mu s$ for store-and-forward switches respectively for a minimum sized packet with 64 bytes payload. For cut-through switches this value is independent

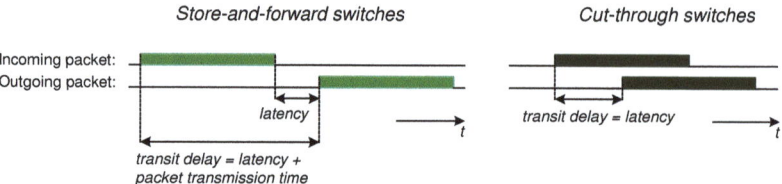

Figure 4.2: Transit delay and latency of switches

of the actual packet size whereas for store-and-forward switches this value is in direct relation to the packet size.

The delay variation added by switches to an end-to-end packet transmission depends on the actual switch type and implemented switching fabric as well as on the load present. Typical values can vary from about 100ns up to some μs as presented in various test results[1] of COTS switches. However, we suggest that this range can be extended up to some ms for stacked switches where a backplane bus is used to cascade several devices.

Ethernet router: An Ethernet router is an intermediate system which operates at the network layer of the OSI model. Routers may be used to connect two or more IP networks (e.g., a LAN with a DSL connection). A router consists of a host processor with at least two network interfaces supporting the IP protocol. The router receives packets via one network interface and forwards the received packets to another one. Received packets have all link layer protocol headers removed, and transmitted packets have a new link protocol header added prior to transmission. The router uses the information held in the network layer header (i.e., the IP header) along with routing information held in a routing table to decide whether to forward a packet or not. Before a packet is forwarded, the processor checks the *Maximum Transfer Unit* (MTU) of the specified interface. Packets larger than the interface's MTU must be fragmented by the router into two or more smaller packets. If a packet is received with the *don't fragment* bit set in the packet header, the packet is not fragmented, but instead discarded. In this case, an ICMP error message is returned to the sender (i.e., to the original packet's IP source address) informing it of the interface's MTU size. This forms the basis for Path MTU discovery.

From the given functionality it is evident that the delays and according variations are at least an order of magnitude greater than for given switching technology. Hence, present routers render tight clock synchronization without additional hardware support impossible.

It is evident that a mechanism to identify the time it takes for every clock synchronization packet to cross an intermediate network element, hub, switch or router is mandatory. Fig. 4.3 illustrates the layered interfaces for the various different Ethernet technologies. The answer to the requirement to add timestamp logic as close as practical to the physical layer is obviously given by the *media independent interface* (MII), see [30, 31, 35] where we first proposed this idea. This interface, present at all Fast-Ethernet controllers, provides a standard interface to couple media access controllers to physical layer devices. Similar

---

[1]The results of several performance tests of COTS switches can be obtained from http://www.veritest.com and http://www.mcclellanconsulting.com.

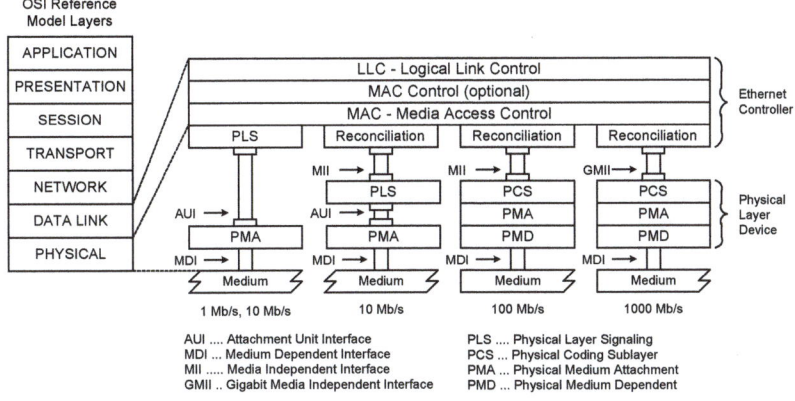

Figure 4.3: Interface structures for different Ethernet technologies

useful interfaces are the *gigabit media independent interface* (GMII) and the evolving new *extended gigabit media independent interface* (XGMII) for Gigabit and 10-Gigabit Ethernet variants.

Timestamping further upstream from the cable —after the media access controller (MAC)— would add the delay variations due to the MAC inherent FIFO's that would limit the synchronization tightness to the μs-range as pointed out in [89]. Timestamping between the physical layer and the cable on the other hand would be very complex because of the required analog interfacing. Furthermore, this is rather impractical since Fast and Gigabit Ethernet employ different codings.

4.2 Network interface for End-systems

This section discusses a network interface architecture for data terminal equipments supporting tight clock synchronization within a distributed system. The presented topologies are based on Fast Ethernet and PCI technologies, which were both chosen because of their dominating and wide spread use in desktop and mobile computers. For both technologies a wealth of implementations and controllers exist. Hence, the presented architecture tries to implement the support for clock synchronization next to existing devices by reusing industry proven technologies rather than to invent new ones. In order to keep the focus on an industrial product, the following goals need to be taken into account:

Functionality: The clock should be implemented in hardware, facilitate clock corrections, support for packet timestamping, enable coupling of external time sources and provide support for applications. It should fit into the standard MII in order to minimize the remote clock reading error. Programming and read-back of the clock registers should be kept straightforward and simple.

Costs: The component count and the size of the printed circuit board should be kept small, therefore standard components should be employed in preference to complex and costly high-end devices. Furthermore, the custom clock chip should employ a small die size and a low pin count.

Transparency: The required support mechanisms should be added so as to allow re-use of all existing communication protocols, hence it should act transparently to all existing mechanisms. Programming and read-back of the clock registers should be accomplished via existing interfaces in order to avoid complex device driver developments. In particular, programming of the clock via a socket interface would ease software developments for various different operating systems.

4.2.1 Clock synchronization support for Network Interface Cards

Tab. 1 in the appendix lists several devices from different vendors that could be employed for a network interface for Fast Ethernet, Gigabit Ethernet or 10 Gigabit Ethernet. Most media access controllers (MAC) support the PCI and MII interface. Some of these devices have an additional built-in CPU and other interfaces as well.

Common to all the different architectures is the media independent interface connecting a media access controller with a physical layer device. In the transmit path this interface consists of the transmit clock ($25MHz$ output from the physical layer device), a transmit enable signal and four transmit data lines. For the receive path the same set of signals exists extended with a carrier sense and a collision detect[2] signal. The receive clock is, as with all asynchronous communication systems, recovered from the incoming data stream and hence not in synchrony with the transmit clock.

Packet oriented Clock Interface

The idea here is to timestamp and program the clock via dedicated *clock synchronization packets* (CSPs). These packets are distinguished by the use of a special type field value. The actual information is transported in the payload at fixed offsets. Fig. 4.4 illustrates this architecture.

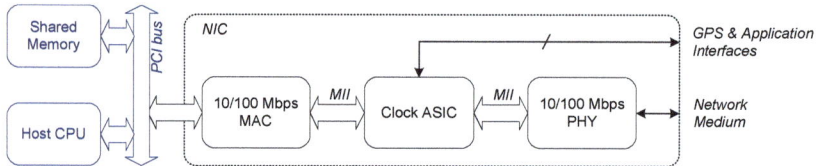

Figure 4.4: Packet Clock Interface

This architecture could go with almost any Fast Ethernet controller and requires very few pins at the clock ASIC. A further advantage is that no custom device driver software is required. Clock synchronization and programming of the clock ASIC could be implemented on top of existing device drivers via a raw socket interface. Hence, software could be easily ported to various different soft+hardware platforms.

---

[2]The carrier sense and the collision detect signals are driven by the physical layer device to indicate whether either data or a collision were encountered on the communication medium. However, all media access controllers used in the experimental evaluation ignore these two signals and derive the same information by simply monitoring the receive and transmit data lines.

Unfortunately, programming and applications usually require a read-back mechanism of the clock ASIC registers. This, in turn, would require the clock ASIC to bounce back a packet that was sent from the host CPU in order to present the register values back to the host.

This mechanism, however, is tricky to implement into the clock ASIC. In half duplex mode a MAC expects the sent data on the MII transmit data lines to be bounced back by the physical layer device onto the receive data lines. Therefore, the clock ASIC must generate a packet in the direction back to the host CPU after the requesting packet has been sent. This, in turn, requires a mechanism to allocate the channel immediately after the interframe gap time of the sent packet has elapsed. In full duplex mode the situation is problematic, too; here a packet could be received in parallel whilst the host CPU requires a read-back of several clock ASIC registers. Either a FIFO for incoming packets needs to be implemented and/or packets are dropped in order to insert the read-back packet into the incoming data stream. Thus, all things considered, a bounce-back mechanism implemented in the clock ASIC requires considerable efforts and may worsen the performance of the network interface since occasionally packets may be dropped.

Alternatively, read-back of registers could be performed by inserting the appropriate register values into received clock synchronization packets. Two possible mechanisms could be implemented to get a timely response. Either the physical layer device is put into loopback mode (this would temporarily disconnect the node), or a remote host is programmed to perform the reply. Use of the latter should be preferred although a mechanism will be required for the case when a host becomes temporarily unavailable. For example, after a suitable timeout, a different remote host should be selected to function as clock read-back partner. Here a way to identify and drop late replies is required. To conserve local time when loss of link is sensed the physical layer should be placed into loopback mode. Now all packets are bounced back by the physical layer device rather than by a remote host. Programming and read-back should still be possible as in the undisturbed scenario.

Packet oriented Clock Interface with split Clock Synchronization Algorithm

The main disadvantage of the packet oriented clock interface is the requirement to read-back clock ASIC registers by the host CPU. This can be improved by implementing the clock synchronization algorithm within one clock ASIC. Most clock synchronization algorithms are round based and require a periodic exchange of clock synchronization packets. Thus a part of the clock synchronization algorithm must be run on the host CPU (the only one that can generate packet send requests via the protocol stack) but the actual clock correction and most other mechanisms could be run on a small CPU next to the clock ASIC, see Fig. 4.5.

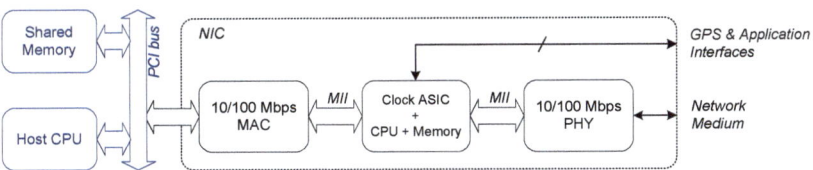

Figure 4.5: Packet Clock Interface with split Clock Synchronization Algorithm

Using this architecture only applications remain that could require access to clock register values of the clock ASIC by the host CPU. This mechanism could be implemented in a similar fashion as illustrated in the previous subsection. Splitting any clock synchronization algorithm accordingly for this particular architecture should be possible without too many problems.

Dedicated Clock Interface

Both solutions mentioned above suffer from the need to exchange information between the clock ASIC and the host CPU. This can be improved by means of a dedicated interface between the clock and the host CPU. Fig. 4.6 illustrates this solution, with the additional data path. Unfortunately only very few devices listed in the Tab. 1 in the appendix sup-

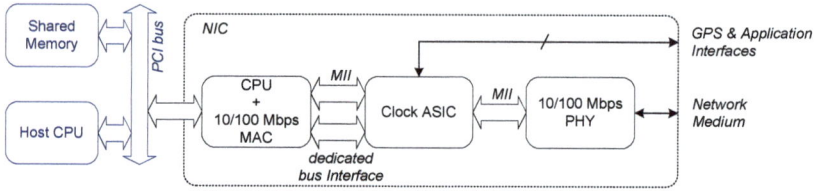

Figure 4.6: Dedicated Clock Interface

port an additional interface, in particular microprocessors with an integrated Fast Ethernet controller like the IBM PowerPC 405GP, the Motorola MPC8265 or the NEC uPD98502. These devices are usually more expensive than dedicated Fast Ethernet MACs but, of course, offer further benefits. They could run the entire clock synchronization algorithm and the network stack that is usually executing on the host CPU. Since popular network stacks like TCP/IP are very computing intensive, such a solution would drastically improve the overall system performance too.

On the other hand, this architecture requires a more costly software since existing device drivers can only be reused up to some extent.

Integrated Clock Interface

The IEEE standard 802.3 restricts the delay of the MII lines between the MAC and the PHY to $7.5ns$ on the same printed circuit board, see [39] Annex 22A. In the previous architectures this timing budget must be fulfilled by the clock ASIC and a corresponding printed circuit board layout; this is a challenging engineering task. In order to overcome this technical problem a device that exhibits a pin-to-pin delay well below this margin or an integration of the clock ASIC with a MAC is advantageous. Fig. 4.7 illustrates this architecture.

The advantages of this architecture are full standard compatibility, direct access of clock registers from the host CPU and higher integration that results in lower production costs for a higher production volume. The main disadvantage of this approach is the requirement of a fully-fledged driver development for every operating system.

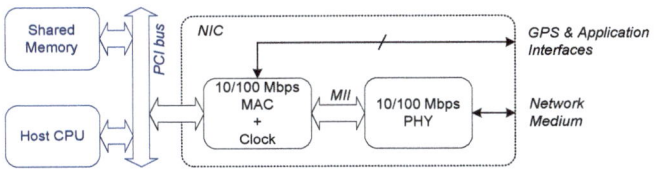

Figure 4.7: Integrated Clock Interface

Contribution to the worst-case transmission delay uncertainty

In all the architectures illustrated above, timestamping of CSP's could be implemented in the same way at the MII, reducing the worst-case transmission delay uncertainty ε. In particular, when the local clock is in synchrony with the MII transmit clock, then a sender simply adds $\varepsilon_{d,trans}^S + G_{ts}^S$, the jitter of the physical layer device in the transmit direction plus the error due to the finite granularity of the timestamp, onto ε. The contribution at the receiver, however, is made up of $\varepsilon_{d,recv}^R + \varepsilon_s + G_{ts}^R$, the jitter of the physical layer device in the receive direction, the additional uncertainty ε_s caused due to a synchronizer stage required to synchronize the received data to the clock domain of the local clock and the timestamp granularity at the receiver. Assuming everywhere the same clock and timestamp logic, i.e. $G_{ts} = G_{ts}^S = G_{ts}^R$, both network interfaces add

$$\varepsilon_{nic} = \varepsilon_{d,trans}^S + \varepsilon_{d,recv}^R + \varepsilon_s + 2G_{ts} \qquad (4.1)$$

onto the transmission delay uncertainty ε. Herein $\varepsilon_s = 1/f_s$ with f_s the sampling frequency used to synchronize the data to the local clock at the receiver. For timestamp's using an NTP-time or similar format covering the bit positions [+m:-n], the error due to the finite granularity is given by $G_{ts} = 2^{-n}$. Assuming that the transmission delay uncertainty of the physical layer devices is approximately the same for both the transmit and receive direction, i.e. $\varepsilon_{d,trans} \approx \varepsilon_{d,recv}$, and that the transmission delay uncertainties of the employed physical layer devices at both endpoints can be bounded by a constant $\varepsilon_d^R \approx \varepsilon_d^S \leq \varepsilon_{dmax}$, we get

$$\varepsilon_{nic} \leq 2\varepsilon_{dmax} + \varepsilon_s + 2G_{ts} \qquad (4.2)$$

4.2.2 Prototype: MII-NTI

In order to prove these concepts a prototype network interface card, termed MII-NTI (Media Independent Interface - Network Time Interface), was developed, see [32]. This prototype printed circuit board, see Fig. 4.8, allows the implementation of the packet oriented clock interface and of the dedicated clock interface, respectively.

A separate PCI-to-PCI bridge is required to comply with the PCI standard. A PCI target device provides the dedicated interface to our custom UTCSU clock ASIC. The functionality of both architectures can be implemented within the FPGA that is placed in-between the MAC and the PHY and the PCI target chip and the UTCSU clock ASIC, respectively.

The basic functionality of timestamping at the MII is illustrated in Fig. 4.9. Clock synchronization packets are recognized by their unique type field (TF). Thus when the timestamp logic at the sender detects this value, when the type field at its fixed offset

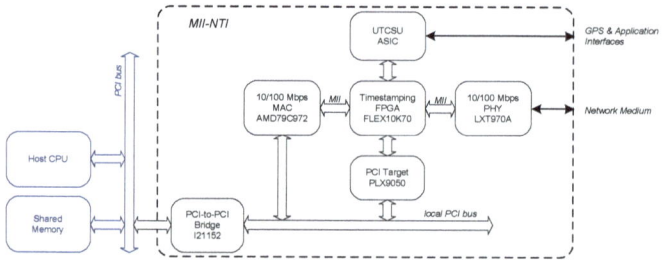

Figure 4.8: MII-NTI prototype architecture

after the preamble is read, it samples a timestamp from the local clock (a) and inserts this value aligned into the packet payload (b), actually overwriting existing data. Since the timestamp value modifies the packet payload, a re-calculation of the frame check sequence (FCS) is required (c) – using the Autodin-II polynomial for Ethernet frames. The newly generated frame check sequence succinctly replaces the old one (d) that was generated and inserted by the MAC.

At the receiver the frame check sequence first needs to be re-calculated starting at the destination address up to the end of the user payload (e). Similar to transmit timestamping the timestamp logic at the receiving node scans the type field value, and on detection of a valid CSP value samples a timestamp from its local clock (f). This timestamp value is then written into the payload of the received frame succinctly after the transmit timestamp (g). —Note that hereby it is required to bit-synchronize the receive data stream to the local clock domain.— Because of the a-new modification of the payload, a re-calculation of the frame check sequence needs to be performed (h) once more. This new value then overwrites the received frame check sequence if no transmit error was detected (i). Otherwise a wrong frame check sequence is inserted to allow for proper detection of the transmission error at the receiver MAC as well.

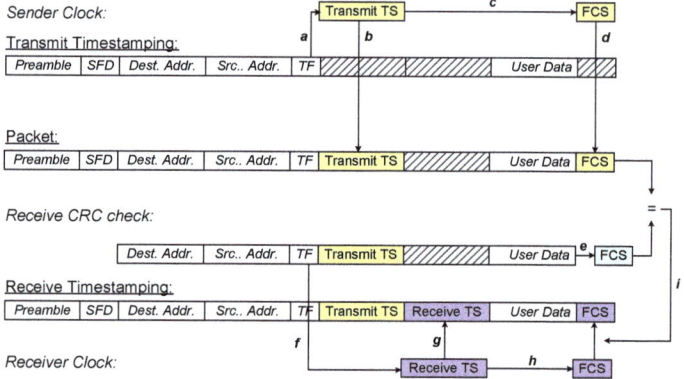

Figure 4.9: MII Timestamping

Programming and read-back of several clock registers in the packet oriented clock interface architecture can be accomplished in the same way. Several bits within the payload field of the clock synchronization packet are interpreted as addresses while others serve as data containers. This mechanism is applicable to the programming of the local clock as well as to the programming of a remote clock, see also Sec. 4.3.1.

4.3 Clock architecture

For tight clock synchronization in the range of $1\mu s$ and below, a dedicated clock circuit is mandatory to maintain and control the progress of the local clock and to support remote clock readings by exact timestamping. Very advanced clock circuits, in this sense, are the pioneering CSU and our UTCSU ASIC, see [97] and Sec. 3. Especially the underlying concepts of the UTCSU ASIC are well suited for this kind of problem. Since the UTCSU was targeted for a precision in the μs-range, it requires some modifications when a precision in the *ns*-range is targeted.

Furthermore, an industrial implementation mandates a small chip with reduced die size and minimal pin count. Size reduction is partly accomplished by different functional implementations, as is explained throughout the next subsections, and by migration to a smaller process technology. The benefit of a smaller process technology also facilitates an increase in the operating frequency, a concern that can have a big influence on the overall clock synchronization, see Sec. 5. The pin count can be drastically reduced by abandoning the multiplexed NTP interface –a 64 bit wide interface used to export the time in a multiplexed fashion– and simplifying the rather complex bus interface.

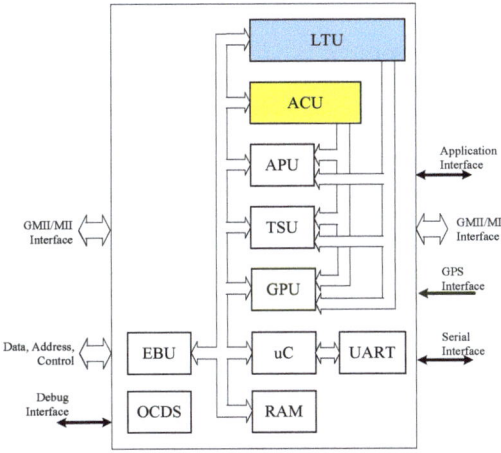

Figure 4.10: Clock ASIC IP-core modules

Fig. 4.10 shows a coarse block diagram of an IP-core solution that resembles several units found in the UTCSU. This solution is built with a set of the following units:

Local Time Unit (LTU): An adder-based clock is used to represent the local clock. For initialization purposes the clock is arbitrarily state-adjustable and allows smooth clock corrections by continuous amortization. To that end, features to switch between pure/amortized clock rates are at hand to commence the amortization phase at a programmable point in time.

Accuracy Unit (ACU): This optional unit maintains accuracy intervals to capture an external reference time. Therefore, these intervals are both adjustable and deteriorate with the progress of time to account for local clock drifts.

Application Unit (APU): In order to support applications requiring access to a precise synchronized distributed clock, polarity programmable input lines allow timestamping of events. A FIFO buffer prevents fast, repetitive events from overwriting the timestamps when event processing is not fast enough.

Timestamp Unit (TSU): Inbound and outbound clock synchronization packets supplied via the MII are recognized by their unique type-field. A timestamp is sampled and on-the-fly inserted into the packet payload at predefined offsets. The frame check sequence of every packet is validated and updated accordingly. A similar technique can be used to set or read the registers with the help of dedicated packets.

GPS Unit (GPU): For access to an external reference timing receiver, this unit timestamps events on a dedicated *1pps* input. The respective status information belonging to the *1pps* signal and typically provided via a serial interface is connected to the UART module.

Micro-Controller (uC): A micro-controller controls the UART module and executes the major part of the clock synchronization algorithm.

External Bus Unit (EBU): A simple non-multiplexed bus interface allows the accessing of every clock register whenever programming via a dedicated interface is implemented.

On-Chip Debug Support (OCDS): A bootstrap and debug facility supports program development.

The following sections propose some mechanisms that should be considered when these modules are actually implemented. An engineered solution is presently developed by the spin-off company Oregano Systems with funding from the Austrian FIT-IT programme.

4.3.1 Bus Interface and Timestamp Unit

Layout and functionality of the clock ASIC registers is designed to support all the previously outlined architectures with different kinds of algorithms. Programming via dedicated network packets provides the benefits that no modification of existing device drivers is required and that the local clock of a node can be directly programmed and controlled by a remote host:

- Clock Programming Packet (CPP): The local host CPU programs the register set of the clock ASIC with this particular network packet identified by a special type field. The payload holds the actual data for every register in advance, either at a pre-defined offset or with an accompanying address.

- Clock Read-Back Packet (CRP): The local host CPU reads the register set of the clock ASIC by sending a clock read-back packet to be bounced back by a remote host. The local clock ASIC recognizes this bounce back packet by its unique type field and inserts the register values into the remaining payload either at predefined offsets or with an address tag in advance. A timeout mechanism at the local host CPU is required to account for situations when a link error occurs. The local host shall switch to a degraded mode when the entire link goes down or select an alternative redundant host when the remote host becomes unavailable.

- Clock Broadcast Packet (CBP): This packet, generated by an elected master node, can be used to set the clock state of remote clocks for initial clock synchronization and/or node join. This packet is identified by a special type field at the start of the packet payload. The register values are found either at a fixed offset within the packet payload or in advance with an address tag. With the knowledge of the network topology, the delays and delay variations one can effectively initialize the distributed clocks; otherwise, this is a difficult task.

- Remote Clock Programming Packet (RCPP): This packet is the same as a CPP but the originator is a remote host rather than the local host. In this way it is possible to correct and control the clock of a remote node. This mechanism becomes especially useful when a node has no processing unit, it's capacity is restricted or when it is simply too expensive to implement a dedicated clock synchronization at this node. To that end a remote node needs to initialize and set all required clock registers via a dedicated programming packet.

- Remote Clock Read-Back Packet (RCRP): The local host responds to a remote request for read-back of the clock ASIC registers. The local host replies with a packet employing a special type field at the start of the packet payload. When the packet is sent, the clock ASIC inserts the requested clock register values transparently into the outgoing packet payload.

For all proposed mechanisms to set or read remote clock registers, a way to identify and authorize the required packets is necessary in order to avoid erroneous behavior caused by "false" packets.

When an implementation is chosen as outlined in Fig. 4.6, 4.7 and 4.8, then direct access via an additional bus interface from a host processor is required.

- Parallel bus interface: This solution provides access to the clock ASIC registers when either a dedicated clock interface or an integrated clock interface architecture is used. All registers can be conveniently programmed/read-back via this interface.

4.3.2 Local Time Unit

The centrepiece of the clock ASIC is formed by the Local Time Unit (LTU) that maintains the local clock supporting continuous amortization. Here phases where the clock is corrected (amortization phase) and phases where it is fly-wheeling (pure phase) interchange perpetually. An adder-based approach is used to carry out fine-granule corrections equidistantly distributed during the amortization phase by adding a programmable small amount to the clock ticks.

The n-bit wide adder maintains the local clock in an extension of the popular NTP format. The meaning of every bit is derived as a power of two — e.g., bit position $+31$ wraps every 68 years while bit position -32 accumulates $2^{-32}s \doteq 233 ps$. Full NTP time consumes 64 bits, where the upper 32 bits are interpreted as standard seconds relative to UTC and the lower 32 bits give the fractional part. For a clock precision down to the ns range additional ultrafractional bits are required. The amount of additional bits is determined as a trade off between limiting factors due to the chip technology in use and the technical parameters for tight synchronization. As stated in Sec. 2.3, the clock precision depends on the parameters:

The remote clock reading error ε: This parameter, employing architectures as illustrated in Sec. 4.2, consists of the transmission jitter of the physical layer (cable + physical layer devices) and the sampling error of the local and remote clock if the local clocks are driven by oscillators not belonging to either the physical transmit or to the receive clock domain: $\varepsilon = \sum \varepsilon_c + \sum \varepsilon_d + \sum \varepsilon_s$. To minimize the effect due to synchronizer stages between different clocking domains we mandate:

- The clock at every node should be driven in synchrony with the transmit clock.

- A high sampling frequency in synchrony with the transmit clock is used to synchronize incoming CSP's and to draw and transparently insert a receive timestamp into the packet (e.g. $100MHz$ add $\varepsilon_s = 10ns$).

The re-synchronization period P and the clock drift ρ: The re-synchronization period P on one hand should be chosen in a way to reduce traffic, on the other hand frequent re-synchronizations keep the clocks tighter synchronized. The clock drift ρ is due to the oscillator characteristics; hence an oscillator should be chosen with characteristics that will not impair the achievable precision in a significant way. Unfortunately most ovenized oscillators with small drift, good stability and reasonable small phase noise characteristics are typically designed for frequencies in the range of $1-20MHz$. Furthermore, MII and GMII require a $25MHz$ and a $125MHz$ clock source, respectively. The "rather low" clock frequency of $20MHz$ however contradicts the above demand for a high sampling frequency, hence a way to increase the operating clock frequency without affecting stability issues is required. One possible solution is the use of either a *phase locked loop* (PLL) or a *delay locked loop* (DLL) for frequency multiplication[3].

---

[3]Note, however, that this has usually worse effects on the oscillator's frequency stability. The phase noise contribution of a PLL depends on the jitter of the oscillator (e.g. a VCO), the jitter of the PLL input source, and the bandwidth of the loop, see [112]. To minimize the effects due to the PLL internals, a VCXO should be used in preference to a VCO as used in MCXO's. This approach is typically taken in COTS frequency translators (FCXO's).

For a DLL the phase noise is primarily due to the noise from the input source and those coupled into the circuit by the delay lines. This effect can be minimized when the amount of delay lines is kept to a minimum, see [11]. Furthermore a DLL may exhibit false locking to multiples of the clock period. In order to avoid this problem, the DLL shall be reset to a minimum delay prior to start-up. Thus, when the DLL attempts to lock, it will increase the delay of the delay line until precisely one period of delay exists in the line, avoiding the false locking.

Other effects, e.g. phase slips, are of no concern for DLL's or PLL's when they have locked as long as the reference oscillator is accurate.

The clock reading granularity G and the rate adjustment uncertainty u: These parameters are directly proportional to $1/f_{osc}$ and become significant when ε becomes very small. Thus these parameters mandate a high clock frequency next to a high sampling frequency of timestamps.

Resolution of the adder: The intrinsic rate R_f of the proposed clock is given by $R_f = G_{sf}/G_t$ where the full state granularity is given by $G_{sf} = 2^{-n}$ and the time granularity by $G_t = 1/f_{osc}$. Suggesting a 64 bit wide adder spanning the bit range [+31,-32] corresponding to the NTP format, and a clock frequency of $f_{osc} = 100MHz$ would yield an intrinsic rate of $R_f = 2^{-32}s/10ns = 23,3*10^{-3}$, which would intolerably impair the rate of every oscillator. Hence, keeping the $100MHz$ one must extend the resolution of the adder up to a suitable fractional bit position. Extending the adder by additional n-bits spanning the range [+31,-n] one gets an intrinsic rate error of $2^{-n}f_{osc}$. The intrinsic rate error should be an order of magnitude below the given oscillator drift that influences the precision by $4P\rho$.

The clock setting granularity G_s directly impairs on the achievable worst-case precision and is given by the lowest bit position of the clock that can be set. When the proposed add-based clock can be programmed down to its lowest bit position the clock setting granularity is determined by $G_s = 2^{-n}$, hence, this effect is negligible for most designs.

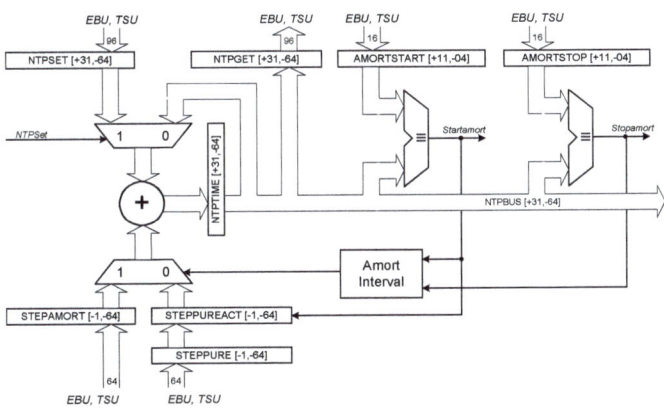

Figure 4.11: Local Time Unit Block-Diagram

Summarizing, an actual implementation should try to optimize both the maximum frequency as well as a suitable resolution. From a practical point of view it makes sense to chose the adder resolution as a multiple of one byte. Fig. 4.11 illustrates a block-diagram of the local time unit, where the adder width is chosen over a range of [+31,-64].

The amortization phase is derived from the local clock controlled with two programmable registers. In particular, the point in time when the amortization phase starts is invoked either periodically or once with the help of register AMORTSTART and a suitable digital comparator. Whenever a certain set of bits programmed into register AMORTSTART is present on the NTPBUS a suitable pulse that invokes the amortization phase is generated. Next to the invocation of the amortization phase, this pulse additionally loads the new value for the following pure phase. The clock value is corrected during amortization phases with the value programmed to register STEPAMORT and during pure phases via STEPPUREACT. The amortization phase ends when the clock value present on the NTPBUS equals the end-value programmed to the register AMORTSTOP.

This mechanism is a fundamental enhancement with respect to the UTCSU, where the duration of the amortization phase was controlled in physical clock ticks rather than logical time as outlined. The disadvantage of using the physical time is the difficulty to determine from any timestamp whether amortization is still in place or not. The lack of this mechanism makes it difficult, for example, to perform correct round-trip measurements.

4.4 Networked devices

Several intermediate networked devices in an end-to-end communications path impair the packet transmission variability as stated in Sec. 4.1. Depending on their type (hub, switch, etc.) and their particular realization (store-and-forward, cut-through, etc.), they add up a certain amount onto the worst case remote clock reading error ε. When these values are intolerable for a certain clock precision, additional measures are required to determine and reduce these values. The following sections illustrate and analyze an architecture that enhances existing switches with hardware support to make tight clock synchronization possible. The underlying ideas can be applied to other devices in a similar fashion.

4.4.1 Clock synchronization support for Switches

A typical block-diagram of a modern, popular switch[4] employing a shared memory architecture is shown in Fig. 4.12.

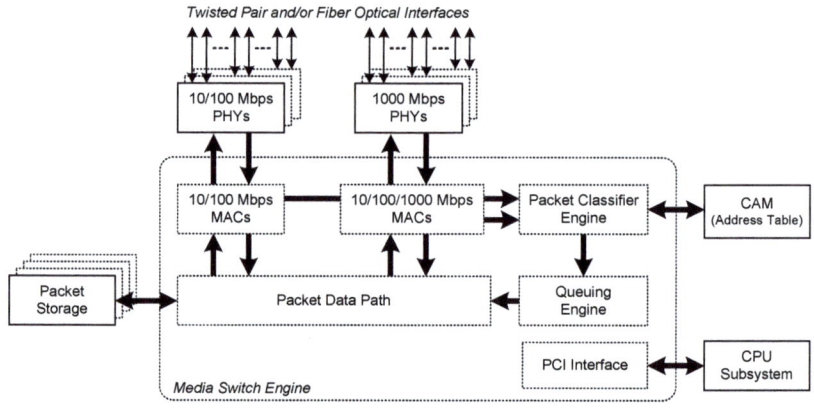

Figure 4.12: A typical modern switch architecture

---

[4]This architecture is a typical application of the Intel Media Switch IXE2424, a 10/100+Gigabit L2/3/4 device, that is used to build cascading high port count Layer 2/3/4 switches or routers.

The main functionality is implemented as a switch fabric that hosts several medium access controllers; one for every interface. A packet classifier engine records the MAC addresses and stores these within a *content addressable memory* (CAM). In addition, it implements a priority control mechanism that provides a *quality-of-service* measure for distinct packet types. The bridging of the packets between the corresponding ports is performed with the help of a queueing engine that is under the control of the packet classifier engine. In case that the required output ports are congested, the packets are buffered in a suitable packet storage. The handling of several modes of operation is usually programmable with the help of a dedicated host CPU. Therefore a standard off-the-shelf PCI interface is integrated into the switch fabric. Just as with all end-systems, the MII or GMII are used to connect every media access controller (MAC) with a physical layer (PHY) unit that provides the interfaces to the switch ports. For example, the IXE2424 switch fabric provides 24 interfaces to Fast Ethernet and 4 interfaces to Gigabit PHY's. One or all of the Gigabit ports can be used to cascade the switch to build a stack of switches.

Since the media independent interface is present within a switch, this interface could be adapted to capture the contribution to the transmission delay uncertainty of every switch. This basic principle, which could be added/integrated to existing hub/switch architectures and behaves transparent to all existing network traffic, is illustrated in Fig. 4.13.

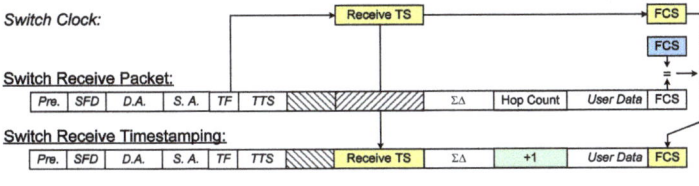

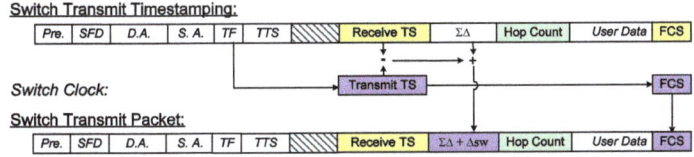

Figure 4.13: Determining the packet delay caused by switches/hubs

It is a similar mechanism as used for timestamping clock synchronization packets; dedicated timestamp logic is placed in the MII path of every hub/switch interface. Whenever a clock synchronization packet is recognized at the receiving MII data lines on any particular interface, a *switch timestamp* is sampled into the packet payload following the fields of transmit and receive timestamps that are inserted at both end-systems. This requires bit-synchronization of the signals from the receive clock domain to that of the local clock domain, respectively. Next to this switch timestamp insertion the frame check sequence of the received, unmodified packet needs to be checked to allow for transmission error detection. When a correct CRC is present, a new CRC calculated off the now modified data needs to be inserted, overwriting the received checksum. In case of a transmission error detection, an incorrect checksum is forwarded to the subsequent MAC unit, to allow for adequate handling of the packet within the switch.

When the packet is forwarded onto a port, a second switch timestamp is sampled off the same clock. The difference to the first timestamp is calculated and added to a further payload field that accumulates all delays that are caused by several switches, hubs and the like in between an end-to-end communications path. Finally, the checksum of the again modified packet needs to be adjusted accordingly. Once more, bit-synchronization is required to transfer the data from the local clock domain to the transmit clock domain[5].

The principle of accumulating all delays due to intermediate systems allows for a fine granular measurement of the variable packet delays with two single timestamp fields within the payload of a dedicated clock synchronization packet. Fig. 4.14 illustrates a switch architecture with the required hardware support for clock synchronization; clocks with timestamp-logic added in the path between the physical layer devices and the switch fabric.

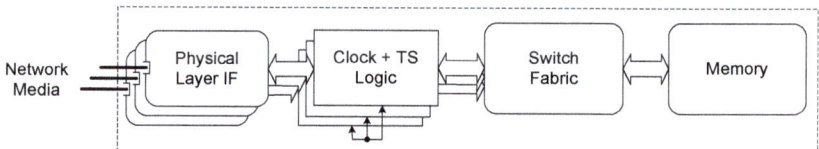

Figure 4.14: Switch architecture with HW-support for clock synchronization

For a product development, this mechanism should be integrated into the switch fabric device, next to the MAC units. Note that a single counter approach for the clock could suffice here when the error due to the oscillator drift is negligible. The latter depends on the clock drift of the oscillator and the transit delay through the switch. When this error becomes significant either a better oscillator or the incorporation of a clock rate mechanism (e.g., an adder-based clock) is required. To provide a clock for timestamping at every switch port different architectures are conceivable as illustrated in Fig. 4.15.

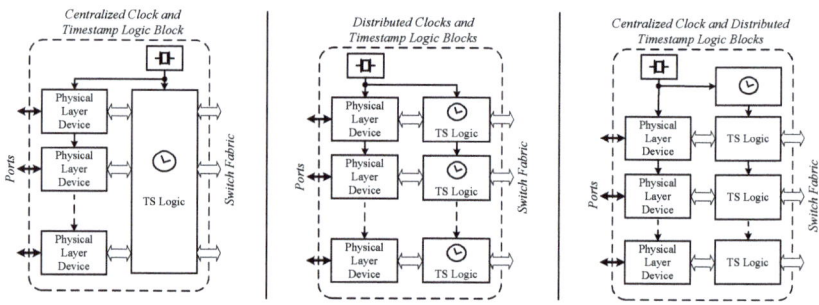

Figure 4.15: Clock Architectures for switches

---

[5]In contrast to the NIC's the local clock used for timestamping at a switch requires an own clocking domain. Since there is a multitude of clocking domains for the receive and transmit paths it doesn't make sense to have the clock domain synchronous to one single such domain.

Centralized clock and timestamp logic

The clock and the timestamp logic for all ports could be implemented with one large ASIC. An adder based clock approach could be used to synchronize the clocks of the switch and to provide a rate correction mechanism which would allow the trading of expensive but precise oven controlled oscillators with ordinary crystal oscillators. The drawbacks of this approach are the poor scalability and the requirement for multiple clocking domains within one chip.

Distributed clock and timestamp logic

By splitting and integrating the logic into several devices one can effectively improve the above mentioned centralized clock and timestamp logic. This approach scales efficiently to the different port counts of switches and eases the handling of the required, multiple clocking domains. Unfortunately, this architecture requires provisions to keep the distributed clocks in lock step — a single bit flip could cause gross errors and would go undetected without additional measures. The lock mechanism of a PLL output or a signature mechanism calculated from the actual clock values and exported via some suitable interface could be used. In the latter case, every device could output its signature onto a Wired-AND bus and read-back the value from the bus in parallel. When a mismatch is detected between the output signature and the signature present on the bus, a re-start of all clocks within the switch could be initiated. Wrong intermediate timestamps should be invalidated in this case.

A synchronous start feature is required for start/re-start of the distributed clocks. The lock output of several PLL circuits used within every clock chip could be used to trigger this mechanism, when all PLL's are locked to the reference clock.

When a rate correction is required, it should be implemented with a microcontroller and a dedicated clock next to the oscillator. The distributed clocks should then operate off the rate synchronized clock.

Centralized clock and distributed timestamp logic

Combining the advantages of both previously sketched architectures, one could distribute the timestamp logic to several devices and implement the clock in one dedicated chip. This solution exhibits good scalability and allows the implementation of a rate algorithm as well. The only drawback here is the wide interface to export the clock values to all timestamp logic devices.

Contribution to the worst-case transmission delay uncertainty

When calculating the transmission delay uncertainty for any of the modified switch architectures one needs to accumulate the jitter due to the physical layer devices, the synchronizer delays, and the influence of the clock drift. The latter is required because of the large transit delay δ_{sw} through a switch which can reach the *ms*-range. For e.g. given a worst case switch transit delay of 200*ms*, a drift of $\rho_{sw} = 10^{-7} s/s$ adds an error of up to 20*ns*. Hence either an oscillator with a drift below $\rho_{sw} = 10^{-8} s/s$, or a suitable clock rate correction mechanism is mandatory.

Given the architecture illustrated in Fig. 4.14 and the assumptions

1. the timestamps at every port are drawn from an identical clock source and

2. identical physical layer devices are used with $\varepsilon_{d,trans} \approx \varepsilon_{d,recv} \leq \varepsilon_d$,

then the maximum transmission delay uncertainty contributed by one single switch is

$$\varepsilon_{sw} = 2\varepsilon_d + 2\varepsilon_s + 2G_{ts} + \delta_{sw}\rho_{sw}. \tag{4.3}$$

For n switches cascaded via usual switch ports the transmission delay uncertainty due to the switches accumulates to

$$\varepsilon_{sw,n} = \sum_{k=1}^{n}(2\varepsilon_{d,k} + 2\varepsilon_{s,k} + 2G_{ts,k} + \delta_{sw,k}\rho_{sw,k}). \tag{4.4}$$

If the switches are all equipped with the same clock and timestamp logic and oscillators and we assume that the jitter due to any physical layer device can be bounded by a constant ε_{dmax}, formally $\varepsilon_{d,k} \leq \varepsilon_{dmax} \quad \forall 1 \leq k \leq n$, Equ. 4.4 can be rewritten as

$$\varepsilon_{sw,n} \leq n(2(\varepsilon_{dmax} + G_{ts} + \varepsilon_s) + \delta_{swmax}\rho_{sw}). \tag{4.5}$$

Herein δ_{swmax} is the worst-case transit delay through one switch, i.e. $\delta_{sw,k} \leq \delta_{swmax} \quad \forall 1 \leq k \leq n$. From here, it follows that we need a hop count as illustrated in Fig. 4.13, since we cannot know otherwise how many switches a clock synchronization packet may have crossed.

When switches are cascaded using some kind of backplane bus (e.g., like the stackable Cisco 3750 series), then equation 4.3 applies for all, however, some means to synchronize the clocks in every switching unit is needed. With the help of some additional signal lines next to the interface used to cascade these devices it should be possible to keep the clocks synchronized and to let them operate in lock-step. Therefore, one should use the distributed clock and timestamp logic architecture preferably since this approach scales and requires only a few signal lines to keep the distributed clock and timestamp logic devices synchronized.

4.4.2 Switch Add-On

To evaluate the above illustrated mechanism, some kind of prototype implementation is required. Since switches are built using the leading edge of technology, a fully fledged prototype implementation is a challenging task. Therefore, in the following sections we present an add-on for standard, off-the-shelf switches that alleviates a proof-of-concept demonstration: One can build some kind of repeater with timestamp logic in front of every switch, see Fig. 4.16. This device has a physical layer on every up/downlink port, where one side is connected to the switch and the other one to the network. The mode of operation (half-duplex, full-duplex, 10 Base-T, 100 Base-Tx) is sensed at the network side by using the auto-sensing capabilities integrated into every physical layer device. Next, one must force the physical layer device towards the switch side into the same operation mode.

The physical clock signal RXCLK is re-generated with a reference oscillator from the packets received on the network side. The received data signals are fed –synchronous to RXCLK– to a clock and timestamp logic which behaves transparent to all packets except for clock synchronization packets. The latter are timestamped and then forwarded to the switch. To allow for timestamping, the RXD data lines need to be bit-synchronized to the clock domain of the clock chip. Furthermore, before the data is sent to the switch once more a bit-synchronization to the transmit clock domain is required.

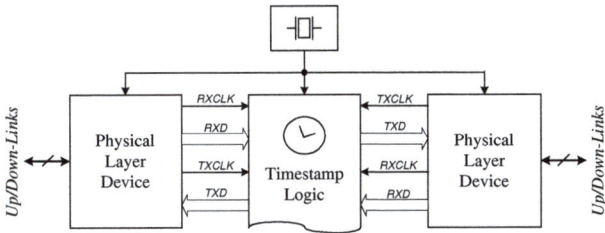

Figure 4.16: Switch Add-On

The same applies in the reverse direction from the switch to the network side whenever a frame is sent by the switch. In this direction, however, the timestamp logic calculates the difference to the timestamp field encapsulated within the packet and adds this difference to a second field within the packet payload. This value accumulates the delays due to all intermediate devices in an end-to-end communications path.

Additional logic –not shown in Fig. 4.16– is required to map the carrier sense and collision detect signals, to handle the auto-sensing capabilities of the physical layer and to provide configuration of all physical layer devices via the bit-serial media independent management data interface that should be controlled via a dedicated, small microcontroller.

Employing this switch add-on architecture, equation 4.3 must be re-written to

$$\varepsilon_{sw} = 4(\varepsilon_d + \varepsilon_s + G_{ts}) + \delta_{sw}\rho_{sw}. \tag{4.6}$$

Similarly equation 4.4 needs to be modified to

$$\varepsilon_{sw,n} = \sum_{k=1}^{n}(4\varepsilon_{d,k} + 4\varepsilon_{s,k} + 4G_{ts,k} + \delta_{sw,k}\rho_{sw,k}). \tag{4.7}$$

4.5 Summary

The presented hardware architecture for tight clock synchronization at DTE devices is centered on a dedicated hardware that provides an adder-based clock, provisions for exact timestamping, and facilities to couple reference timing receivers. This clock is best integrated into the media independent interface between COTS media access controllers and physical layer devices. The same clock technology could be re-used for intermediate networked devices (hubs, switches, ...); however, a simple counter clock with the according timestamp logic could suffice here, given that the error due to the oscillator drift is typically negligible. Several different possible realizations are sketched and analyzed under the aspect of minimizing the remote clock reading error under given implementation specific constraints. For any implementation with n switches in an end-to-end communications path using one of the proposed architectures at every node the worst-case transmission delay uncertainty is computed from

$$\varepsilon = \varepsilon_{nic} + \sum_{k=1}^{n}\varepsilon_{sw,k} + \sum_{k=1}^{n-1}\varepsilon_{c,k} \tag{4.8}$$

where $\varepsilon_{c,k}$ accounts for the individual transmission delay uncertainties caused by the employed cable segments. Assuming near identical physical layer devices at every node with approximately the same jitter for both the transmit and receive direction $\varepsilon_{d,trans} \approx \varepsilon_{d,recv}$, and the use of identical clock + timestamp logic and oscillators built into every switch we can derive from Equ. 4.2 and Equ. 4.5

$$\varepsilon \leq 2(n+1)(\varepsilon_{dmax} + G_{ts}) + \frac{2n+1}{f_s} + n\delta_{swmax}\rho_{sw} + \sum_{k=1}^{n+1} \varepsilon_{c,k}. \qquad (4.9)$$

When standard UTP cables are used and the transmission lines are properly terminated at every node, the jitter due to the cable segments will be negligibly small. Hence, the transmission delay uncertainty is approximately given by

$$\varepsilon \sim 2(n+1)(\varepsilon_{dmax} + G_{ts}) + \frac{2n+1}{f_s} + n\delta_{swmax}\rho_{sw}. \qquad (4.10)$$

Chapter 5
Delay variations of the Physical Layer

The architectural concepts of the previous section were proposed to improve the achievable precision for clock synchronization in a distributed system given by Equ. 2.3. In order to judge the improvement an estimate of the underlying parameters is required. In an implementation ρ, G and u are determined by the oscillator characteristics and the clock design. More important, the clock reading error ε for the presented architectures is made up by two factors:

1. The errors due to synchronizer stages between the different clocking domains of the media independent interface and the clock chip.

2. The delay variation of the underlying physical layer, in particular due to the jitter caused by the physical layer devices and the cable.

The influence of the synchronizer stages can be reduced when a higher frequency for the clock chip is chosen. Unfortunately, a higher clock frequency usually means reduced oscillator characteristics (drift, stability, etc.). At any rate, these parameters can be chosen at design time.

The remaining chapter presents an experimental evaluation of the delay variations due to the physical layer in different end-to-end communication paths for Ethernet variants based on structured copper cabling (10 Base-T, 100 Base-Tx and 1000 Base-Tx). To that end, a model of the cable and the physical layer devices is given before the measurement setup for the different experiments is described. Finally, the measurement results are presented and analyzed.

5.1 Models of the physical communication link

Every end-to-end communications path based on twisted pair Ethernet technologies can be de-composed into several segments (e.g., end-system to switch, switch to router, etc.). Equipped with the hardware support outlined in the previous chapter, the delay variations of the communications path in one way as illustrated in Fig. 5.1 adds up to the remote clock reading error, see Equ. 4.8.

5.1.1 Cable model

For 10 Base-T, 100 Base-Tx and 1000 Base-Tx, a balanced twisted-pair transmission line is employed. Like any electromagnetic transmission line, its characteristic impedance

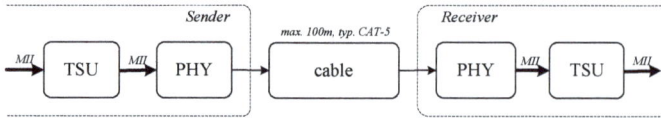

Figure 5.1: Model of the physical communication link (MII-to-MII)

Z_0 can be calculated from manufacturers data and measured on an instrument such as the Agilent 4395A network analyzer. Twisted-pair lines for LAN applications are typically fashioned from #22 AWG or #24 AWG stranded copper wire and categorized for a maximum bandwidth. Category 5 (CAT-5) cables specified for a maximum bandwidth of $BW_{max} < 100MHz$ have tightly twisted pairs for low crosstalk and are often used for 10 Base-T or 100 Base-Tx networks. Category 6 and 7 cables feature a higher bandwidth (350 and 600 MHz) and are in use for Gigabit Ethernet and the like.

The cable from transmission-line theory can be modelled by breaking the line into small parts so that the circuit element dimensions will be much smaller than a wavelength ($\Delta z \to 0$). Doing this, one can describe the transmission-line by a series resistance R $[\Omega/m]$, series inductance L $[H/m]$, shunt conductance G $[S/m]$ and shunt capacitance C $[F/m]$ per unit length. A small section of the transmission-line with length dz thus has the equivalent circuit as illustrated in Fig. 5.2. Analysis of this circuit using Kirchhoff's laws

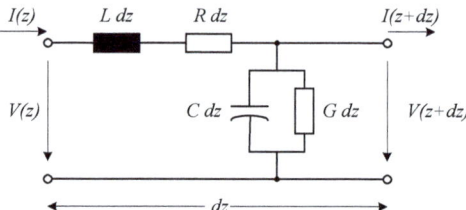

Figure 5.2: Equivalent circuit for a small part of a transmission line

for time-harmonic signals gives the wave equations

$$\frac{\partial V(z)}{\partial z} - \gamma^2 V(z) = 0 \qquad \frac{\partial I(z)}{\partial z} - \gamma^2 I(z) = 0$$

where γ is the complex propagation constant given by

$$\gamma = \alpha + j\beta = \sqrt{(R' + j\omega L')(G' + j\omega C')}. \tag{5.1}$$

$$R' = \frac{\partial R}{\partial z}, \qquad L' = \frac{\partial L}{\partial z}, \qquad C' = \frac{\partial C}{\partial z}, \qquad G' = \frac{\partial G}{\partial z}$$

are the longitudinal derivatives of resistance, inductance, capacitance, and admittance along the line. The solutions to the wave equations are superpositions of forward and reverse waves,

$$V(z) = V_0^+ e^{-\gamma z} + V_0^- e^{\gamma z}, \qquad I(z) = I_0^+ e^{-\gamma z} + I_0^- e^{\gamma z}.$$

The characteristic impedance, defined as the ratio of voltage to current (for positive travelling waves), gives

$$Z_0 = \frac{V_0^+}{I_0^+} = -\frac{V_0^-}{I_0^-} = \sqrt{\frac{R' + j\omega L'}{G' + j\omega C'}}.$$

Furthermore, the phase velocity and wavelength are given by

$$v_p = \frac{\omega}{\beta} = f\lambda \qquad \lambda = \frac{2\pi}{\beta}. \tag{5.2}$$

Combining Equ. 5.1 with Equ. 5.2 one can see that the phase velocity and thus the propagation delay depends on the longitudinal derivatives and the frequency of the time harmonic input signal. As long as the derivatives and the frequency stay constant, the cable adds no delay variation onto the transmission.

In the context of a digital communications link, jitter is the offset between the expected position of a signal transition and the actual position of the transition. Two types of jitter are characterized: *deterministic jitter* and *random jitter*. Deterministic jitter is generally bounded in amplitude, non-Gaussian and expressed in units of time, peak to peak. Examples of deterministic jitter are: Intersymbol Interference (e.g., from channel dispersion or filtering), reflections, duty-cycle distortion (e.g. from asymmetric rise/fall times) and uncorrelated jitter (e.g. from crosstalk by other signals). Random jitter is assumed to be Gaussian in nature and accumulates from thermal noise sources, e.g., small changes of the derivatives due to environmental influences.

For Ethernet systems using twisted pair cabling, the main sources of propagation delay jitter of the cabling are due the output driver at the sender and the input circuitry at the receiver (see also Sec. 5.1.3).

5.1.2 10 Base-T Physical Layer Devices

Fig. 5.3 shows a block diagram that resembles a typical 10Base-T physical layer device. The transmit path consists of a parallel to serial conversion logic followed by a Manchester encoder, a filter and a differential driver. Manchester code is a self-clocking code with a minimum of one and a maximum of two level transitions per bit. A Zero is encoded as a Low-to-High transition, a One is encoded as a High-to-Low transition. The encoded bitstream or link pulses, which are used for auto-negotiation, are feed via a pulse shaping filter to a differential output driver. Link pulses are transmitted every *16ms* in the absence of transmitted data and are used to check the integrity of the connection with the remote end. If valid link pulses are not received, the link detector disables the 10Base-T twisted pair transmitter, receiver and collision detection functions.

In the reverse direction, a differential input buffer feeds the received signals to

- a link pulse detection logic that performs the auto-negotiation functionality,

- to a collision detect logic that reports the presence of a collision in half-duplex mode when data are received and transmitted simultaneously,

- to a smart squelch that employs a combination of amplitude and timing measurements to determine the validity of data on the twisted pair inputs,

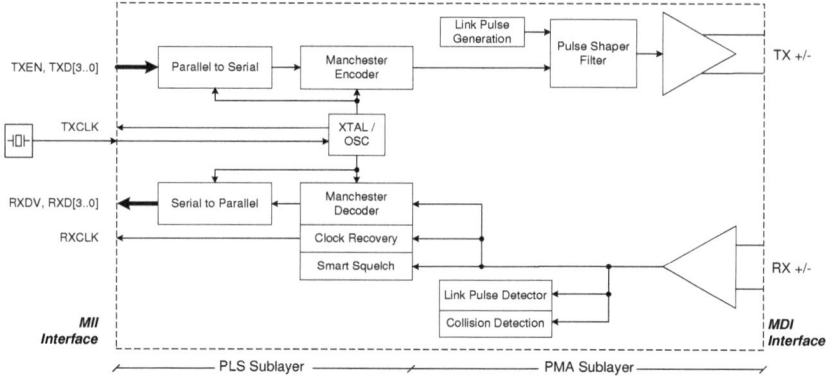

Figure 5.3: A typical 10Mb/s physical layer interface block diagram

- to a clock recovery module that regenerates the MII receive clock out of the received datastream with the help of the local oscillator,
- and finally to a Manchester decoder that uses the recovered clock signal to decode the bitstream.

The recovered bitstream is succinctly serial to parallel converted and output to the MII interface.

Intel LXT970A	
Transmit Latency	typ. $300-500ns$
Receive Latency	max. $7.3\mu s$
LSI-Logic L80225	
Transmit Latency	max. 600ns
Transmit Output Jitter	$\pm 5.5ns$
Receive Latency	max. $3.6\mu s$
Receive Input Jitter	$\pm 13.5ns$
National Semiconductor DP83847A	
Transmit Latency	max. $680ns$
Receive Latency	max. $1.73\mu s$

Table 5.1: Propagation delays and jitter of commercial PHYs

Tab. 5.1 summarizes the relevant delays and delay variations that are relevant for the clock reading error as presented in datasheets of some COTS 10/100Mbps Fast-Ethernet physical layer devices for 10 Base-T MII operation.

5.1.3 100 Base-Tx Physical Layer Devices

Because of the higher signalling rate in Fast/Gigabit and 10Gigabit Ethernet signal dispersion, signal attenuation and electromagnetic emission are a major concern in the physical medium. Furthermore, delay skew between different pairs in parallel transmissions, that is allowed to be in the range of up to 50*ns*, and other effects need to be considered as well. Hence these systems require different line codings as well as a far more complex physical layer interface.

Fig. 5.4 illustrates the physical layer block diagram for 100 Base-Tx and 100 Base-Fx as well as for one channel of 100 Base-T4.

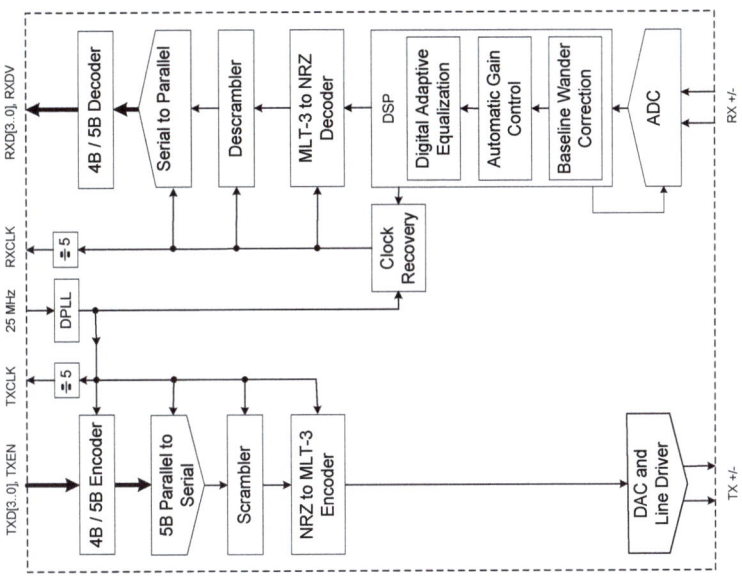

Figure 5.4: A typical 100Mb/s physical layer interface block diagram

In the transmit block a 4B/5B Encoder converts 4-bit nibble data, generated by the MAC and provided via the MII, into 5-bit *non return to zero* (NRZ) code groups for transmission. This conversion is required for control data to be combined with packet data code groups. Certain control code groups replace the preamble and others are appended to the end of a frame; in addition, so called idle code groups are continuously injected into the transmit data stream after a packet until the next transmit packet is detected. Next, the code groups are serialized and fed into a scrambler. Here a closed loop linear feedback shift register with an 11-bit polynomial is used. Scrambling of the data is required in order to randomly distribute the total energy launched onto the cable over a wide frequency spectrum. After the data stream has been serialized and scrambled, the data is NRZI encoded in order to comply with the standard for 100Base-Tx. Next, a binary to MLT-3 conversion is accomplished that outputs two binary data streams with alternately phased logic one events. These two binary streams are then fed to a digital analog converter and a differential output driver. MLT-3 has three signal levels +1, 0 and -1. A logical 1 is represented as a transition from one level to the next, while a logical 0 is represented by no transition. All transitions must follow the repeated pattern: 0, +1, 0, -1.

The receive path consists of a differential input buffer and a fast *flash type analog/digital-converter* (ADC) at the side close to the medium. A differential interface off the twisted pair (electrically decoupled with the help of external magnetics module) provides the input to this ADC. A digital signal processing unit provides baseline wander correction, automatic gain control and adaptive equalization:

- The *baseline wander* (BLW) correction circuit deals with the change in the average DC content, over time, of an AC coupled digital transmission over a given transmission medium. Baseline wander results from the interaction between the low frequency components of a transmitted bit stream and the frequency response of the AC coupling components within the transmission system. If the low frequency content of the digital bit stream goes below the low frequency pole of the AC coupling transformers then the droop characteristics of the transformers will dominate, resulting in potentially serious BLW, see Fig. 5.5.

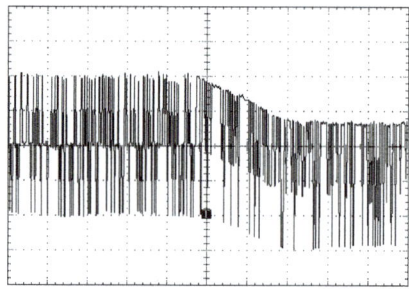

Figure 5.5: Baseline wander

- An automatic gain control corrects signal attenuation and an adaptive equalization filter provides an estimate for the transmission channel. This is required since in high-speed twisted pair signalling, the frequency content of the transmitted signal can vary greatly during normal operation based primarily on the randomness of the scrambled data stream. This variation in signal attenuation caused by frequency variations must be compensated for to ensure the integrity of the transmission, see Fig. 5.6 for a typical signal attenuation/dispersion scenario over CAT-5 twisted pair cabling.

- In order to ensure quality transmission when employing MLT-3 encoding, the compensation must be able to adapt to various cable lengths and cable types depending on the installed environment. The selection of long cable lengths for a given implementation requires significant compensation which will over-compensate for shorter, less attenuating lengths. Conversely, the selection of short or intermediate cable lengths requiring less compensation will cause serious under-compensation for longer length cables. Therefore, the compensation or equalization must be adaptive to ensure proper conditioning of the received signal independent of the cable length. Many designs[1] use an adaptive equalization scheme that determines the approximate cable length by monitoring signal attenuation at certain frequencies. This

---

[1] The information provided herein is excerpted from various data-sheets of different vendors, e.g. AMD, Intel, SmSC, LSI-Logic, National Semiconductor and 3COM.

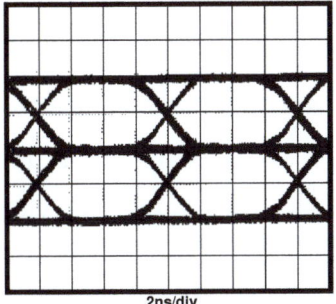 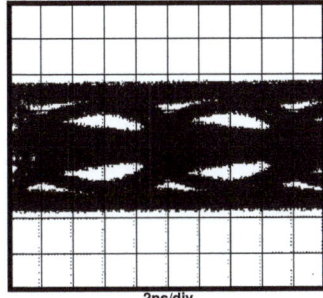

Figure 5.6: MLT-3 dispersion and attenuation on CAT-5 after 0m (left) and 50m (right) measured at the active input interface

attenuation value is compared to the internal receive input reference voltage. The result indicates the amount of equalization to use. Next, the Digital Equalizer removes *Inter Symbol Interference* (ISI) from the receive data stream by adapting a filter to the inverse frequency response of the channel. In combination with a gain stage it is possible to open the receive *eye pattern* sufficiently for reliable data recovery. Some implemented adaptive equalizers select 1 of N filters in an attempt to match the cable characteristics, while others are truly adaptive. Usually, the cable length is estimated based on comparisons of incoming signal strength against some known cable characteristics. The equalizer tunes itself automatically to the cable length to compensate for the amplitude and phase distortion incurred by the cable. The curves given in Fig. 5.7 illustrate attenuation at certain frequencies for given cable lengths. This is derived from the worst case frequency vs. attenuation figures as specified in the EIA/TIA Bulletin TSB-36. These curves indicate the significant variations in signal attenuation that must be compensated for by the receive adaptive equalization circuit.

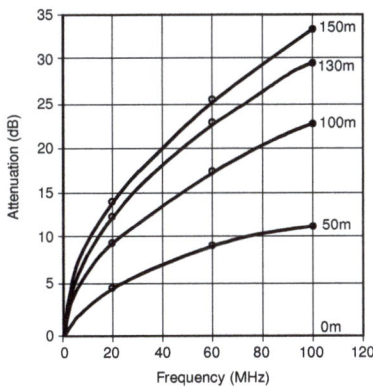

Figure 5.7: EIA/TIA Attenuation vs. Frequency of CAT-5 Cable

Following the adaptive equalization filter are several decoding stages, the clock recovery circuit, a descrambler and a serial-to-parallel converter.

In 100 Base-T4 systems, data is transmitted in parallel over four differential UTP-3 pairs, hence an additional code group align block is present for these systems. Since the number of twists between the pairs in a cable vary, a delay skew of up to $50ns$ is allowed. The code group align block re-aligns the signals receipt on the different pairs and provides them aligned to the serial-to-parallel converter. The latter provides 5 symbols in parallel to the 4B/5B decoder that feeds the nibble wide MII receive data lines synchronously with every rising edge of the receive clock re-generated by the clock recovery circuit.

In 100 Base-T2 systems data is simultaneously transmitted in a full-duplex fashion over two wire UTP-3 pairs in both signalling directions. This dual duplex transmission requires two transmitters and two receivers at each end of a link, and separation of simultaneously transmitted and received signals on each wire pair. To allow for proper operation additional adaptive digital filters are required for echo and NEXT cancellation, equalization and interference suppression, see [10]. These techniques are similar in 1000 Base-T networks.

Fiber optic media are far superior concerning signal attenuation when compared with copper cables. Therefore, the rather complex operations required for signal regeneration, in particular adaptive equalization and baseline wander correction are not required for 100 Base-Fx.

Tab. 5.2 summarizes the relevant delays and delay variations that are relevant for the clock reading error as presented in datasheets of some COTS 10/100Mbps Fast-Ethernet physical layer devices for 100 Base-Tx MII operation.

Intel LXT970A	
Transmit Latency	$40 - 60ns$
Receive Latency	typ. $20ns$
LSI-Logic L80225	
Transmit Latency	$60 - 140ns$
Transmit Output Jitter	$\pm 0.7ns$
Receive Latency	max. $240ns$
Receive Input Jitter	$\pm 3ns$
National Semiconductor DP83847A	
Transmit Latency	max. $60ns$
Transmit Output Jitter	max. $1.4ns$
Receive Latency	max. $21ns$

Table 5.2: Propagation delays and jitter of commercial PHYs

5.2 Evaluation

By experimental evaluation we want to quantify the clock reading error with typical COTS components and the minimal delay variation added due to intermediate switches and repeaters. This is of utmost importance, since side-effects that may impair the system performance can remain unnoticed because of the complexity of the involved mechanisms.

As a result we expect to gain some insights about the jitter added by the complex physical layer devices in order to further improve on the achievable clock reading error. First, the evaluation system consisting of hard- and software is illustrated and described. Next, we characterize the measurement setup and identify key parameters that need to be addressed before we present the measurement results with a proper discussion.

5.2.1 Evaluation System Hardware

Fig. 5.8 illustrates the measurement hardware setup used for the presented experiments. The basic setup remains the same for different *unit under test* configurations and devices.

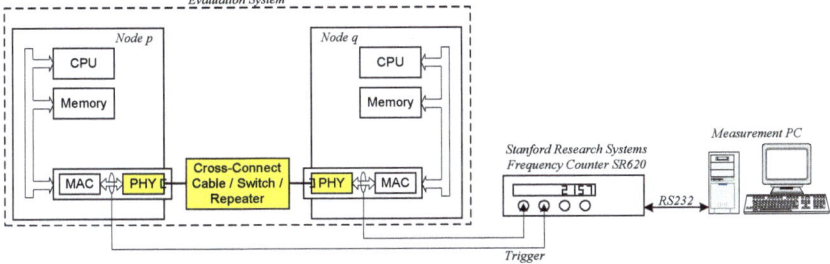

Figure 5.8: Measurement System

Two nodes p and q are equipped with a network interface card that hosts a medium access controller (MAC) coupled via a media-independent interface (MII) to a physical layer device (PHY). The physical layer devices at both nodes and the respective network connection form the unit under test (highlighted in Fig. 5.8). Additional network interface cards (not illustrated) at both nodes provide access to an NTP time server to synchronize their PC clocks for coordinated action. A Stanford Research Systems SR620 Frequency Counter[2] with a RMS resolution of $25\,ps$ and an ovenized timebase are used in conjunction with a measurement PC to record latency variations in packet transmission. Every single measurement is triggered by the rising edge of the transmit enable signal and stopped by the rising edge of the receive data valid signal at either MII. Both signals are synchronous to their according data lines; hence this measurement yields the same result as when triggered and stopped after detection of a unique type field as used for hardware timestamping of clock synchronization packets without the synchronizer stages. Setup and measurement recording are controlled from a remote host via a standard serial interface (RS232). The measurement PC also serves as data repository for the accumulated data.

5.2.2 Evaluation System Software

The software used for the system evaluation consists of

1. a traffic generator for CSP generation

2. a program to control and read-out the frequency counter from a remote host

---
[2]http://www.srsys.com/

3. a program to preprocess the data for graphical presentation

4. and a statistics software package

First, a simple program based on the link-layer library LibNet[3] was developed for low-level traffic generation. This program allows setting the sizes and values of all fields of an Ethernet frame. Additionally, one can program a time interval elapsing between subsequent packets and the amount of packets that are generated. One should note, however, that when executing the traffic generator under a Multitasking Operating System (e.g., Linux, Unix, etc.) the actual time intervals between subsequent packets will vary. This can be alleviated, when

- the system isn't loaded and the traffic generator program is *re-niced* to the highest priority, or when

- a Real-Time Operating System is used (e.g., VxWorks, OSE, RTLinux, etc.).

Throughout the experiments the traffic generation software was executed under the Linux operating system with higher than user priority running atop standard PCs (the nodes of the evaluation system).

The custom developed remote control software was executed from the measurement PC, running under Linux as well. Using the serial interface, this program allows to configure and control every function of the frequency counter and permits a readout of the measurement data. The latter is stored in ASCII files for convenient post processing and analysis. Although the frequency counter could perform up to 10^6 time interval measurements between any two channels and process some statistics (mean, minimum and maximum), it was decided to use single time interval measurements instead. The advantage is that every single data value is available for further post-processing and analysis. In fact, this proved essential for the 10 Base-T and 1000 Base-Tx measurements where the distribution of the data is not Normal.

For every configuration more than 100.000 packets were recorded to provide meaningful output values. To provide a graphic histogram of the distribution the data was pre-processed with a simple, custom developed program that accepts as input the number of histogram-bars the final output shall contain. The resulting values can be plotted using the public domain software Gnuplot[4].

Standard statistical tests like averages, standard deviation and minima/maxima are computed using the public domain statistic program Statist[5]. Furthermore, for characterization of long tails and rare events 95% minima/maxima are provided as well.

5.2.3 Evaluation System Setup

Cables, network interface cards of different manufacturers and repeaters and switches of different brands were subject of the investigation. Therefore, every single experiment was executed in the following sequence:

1. Assembly and setup of the units under test

2. Start-up of the nodes and the measurement equipment

---

[3] http://www.packetfactory.net
[4] http://www.gnuplot.info/
[5] http://www.usf.uni-osnabrueck.de/~breiter/tools/statist/index.en.html

3. Remote initialization of the frequency counter and start of the measurements

4. Forcing of the media speeds and operating modes

5. Invocation of the traffic generator program

6. Data recording

For Fast-Ethernet devices media speed (10Base-T and 100Base-Tx) and mode (half- and full-duplex) can be forced using the Linux system program `mii-tool` that is part of the net-tools[6] package. The latter is usually present in every Linux distribution since it provides other essential networking programs like `ifconfig`, `arp`, `route`, etc.

For Gigabit devices, however, no such handy tool exists; instead, one must rely on a kernel module to provide this configuration mechanism via suitable parameters. First, one needs to off-load the module and subsequently re-load it using either the `insmod` or the `modprobe` system commands. The actual syntax is usually provided by the module documentation or must be grabbed from the module source itself. —A bug was encountered in one version of a Linux kernel module for the Gigabit Ethernet cards employed, preventing proper mode and media speed switching. A bug-fix was was made and sent to Donald Becker[7], the maintainer of the Linux device driver.— In order to verify the media settings one should check the mode and media speed LED's connected to the proper physical layer device and measure the timing and activity on the MII for a packet with known length.

Prior to our experiments we explored some key parameters and their influence on the transmission delays. The following Tab. 5.3 lists them along with a rule-of-thumb characterization of their actual effect on the clock reading error.

Parameter	Effect
Network load	relevant
Loss of link (EMC, disconnection, etc.)	relevant
Length of network segment	relevant
Physical Layer Device	relevant
CPU and interrupt load	irrelevant
CSP size/frequency	irrelevant
Packet loss	irrelevant
Collisions/Late collisions	irrelevant

Table 5.3: Parameters potentially affecting the clock readings error

Gluing all the hardware and the software together following the above steps, different configurations of the physical layer connection in different modes were evaluated:

- Cross-connect cables with different length

- Different COTS physical layer devices

    - forced to media speeds: 10Base-T, 100Base-Tx and 1000Base-Tx

    - configured to the operating modes: Full-Duplex, Half-Duplex

---

[6]http://freshmeat.net/
[7]http://www.scyld.com

- Loss-of-link

- With Unloaded/loaded switches/repeaters in the end-to-end communications path using different media speeds and operating modes

5.3 Measurement Results

The measurement results are presented based on the items listed before. First, the delay variations of up to 100.000 clock synchronization packets employing different physical layer devices and cable lengths are summed up. The results will give realistic values for the clock reading error from which one can conclude whether the particular media speed and mode configuration will admit a synchronization precision in the *ns*-range.

Next, to justify the additional hardware support for switches, we present some test results with switches and some repeaters, respectively. The experiments conducted for both unloaded and loaded conditions are backed-up by other switch evaluations with advanced measurement equipment found in respective literature.

5.3.1 Direct connection

Tab. 5.4 lists the physical layer devices and cross-connect cables that were evaluated. One custom-developed prototype, as illustrated in Fig. 4.8 and several COTS network interface card were employed. The latter were patched so that measurement of the MII control and data lines could be conveniently done. For either physical layer interface two cards were

Network Interface Card	Physical Layer Device
MII-NTI prototype	Intel LXT970A
3Com 3C905Tx	National Semiconductor DP83840A
Allied Telesyn AT2700Tx	Intel LXT970A
D-Link DFE530Tx	LSI Logic L80225
D-Link DGE500T (Gigabit)	National Semiconductor DP83861
Vendor	**Cable**
YFC Boneagle	3m CAT-5 100MHz 26 AWGx4P
Draka Comteq	99m UC300 S26 4P Category-5e

Table 5.4: Physical layer devices and cables, subjects of the evaluation

available (except for the D-Link DFE-530Tx card) to explore send and receive operation with the same devices within one single experiment as well. All experiments were conducted with a 3m and a 99m cross-connected CAT-5 cable to investigate the influence of the cable length. Note that twisted pair Ethernet specifies a maximum segment length of 100m. To that end the experiments with the 99m cable could cover proper operation close to the specified boundaries too.

The results of the experiments are categorized into 10 Base-T, 100 Base-Tx and 1000 Base-Tx operating modes. Figures of the distributions found are given for some typical configurations only and are omitted for other configurations, when leading to the same principal results. Benchmark data (minima/maxima, mean, etc.) are given for all measurements in tabular form to allow convenient comparison and analysis.

10 Base-T results: Tab. 5.5 summarizes the measured delays of 100.000 clock synchronization packets for various Fast-Ethernet network interfaces and two cable lengths in 10 Base-T mode. For every experiment the source, the cross-connect cable and the destination NIC are listed followed by extracted intervals and the samples contained therein. For the given configurations the results are distributed over several intervals spaced about 100ns from each other, see Fig. 5.9 for a representative sample plot of one experiment sequence and an extracted histogram.

Full Duplex		Half Duplex	
Src: 3C905Tx — 3m CAT-5 — Dst: 3C905Tx			
$[2.1110\mu s, 2.1150\mu s]$	10713 samples	$[2.1111\mu s, 2.1151\mu s]$	10903 samples
$[2.2106\mu s, 2.2136\mu s]$	22986 samples	$[2.2107\mu s, 2.2137\mu s]$	23381 samples
$[2.3102\mu s, 2.3122\mu s]$	24925 samples	$[2.3103\mu s, 2.3123\mu s]$	24750 samples
$[2.4103\mu s, 2.4118\mu s]$	25251 samples	$[2.4104\mu s, 2.4119\mu s]$	25013 samples
$[2.5094\mu s, 2.5119\mu s]$	14508 samples	$[2.5095\mu s, 2.5125\mu s]$	14285 samples
$[2.6100\mu s, 2.6115\mu s]$	1617 samples	$[2.6100\mu s, 2.6115\mu s]$	1668 samples
Src: 3C905Tx — 99m CAT-5 — Dst: 3C905Tx			
$[2.6213\mu s, 2.6253\mu s]$	10666 samples	$[2.6215\mu s, 2.6250\mu s]$	10565 samples
$[2.7211\mu s, 2.7241\mu s]$	23252 samples	$[2.7212\mu s, 2.7243\mu s]$	23223 samples
$[2.8209\mu s, 2.8229\mu s]$	24774 samples	$[2.8210\mu s, 2.8230\mu s]$	24977 samples
$[2.9206\mu s, 2.9221\mu s]$	25225 samples	$[2.9208\mu s, 2.9223\mu s]$	25208 samples
$[3.0199\mu s, 3.0224\mu s]$	14380 samples	$[3.0200\mu s, 3.0220\mu s]$	14336 samples
$[3.1207\mu s, 3.1227\mu s]$	1703 samples	$[3.1208\mu s, 3.1228\mu s]$	1690 samples
Src: 3C905Tx — 3m CAT-5 — Dst: DFE530Tx			
$[3.1089\mu s, 3.1105\mu s]$	14824 samples	$[3.1090\mu s, 3.1106\mu s]$	15194 samples
$[3.2089\mu s, 3.2105\mu s]$	25112 samples	$[3.2091\mu s, 3.2107\mu s]$	25047 samples
$[3.3089\mu s, 3.3105\mu s]$	25043 samples	$[3.3087\mu s, 3.3103\mu s]$	25090 samples
$[3.4089\mu s, 3.4105\mu s]$	25228 samples	$[3.4092\mu s, 3.4104\mu s]$	24836 samples
$[3.5089\mu s, 3.5106\mu s]$	9793 samples	$[3.5092\mu s, 3.5108\mu s]$	9833 samples
Src: 3C905Tx — 99m CAT-5 — Dst: DFE530Tx			
$[3.6113\mu s, 3.6138\mu s]$	14362 samples	$[3.6114\mu s, 3.6130\mu s]$	14339 samples
$[3.7113\mu s, 3.7129\mu s]$	24845 samples	$[3.7114\mu s, 3.7131\mu s]$	24628 samples
$[3.8113\mu s, 3.8125\mu s]$	25123 samples	$[3.8115\mu s, 3.8135\mu s]$	25165 samples
$[3.9113\mu s, 3.9129\mu s]$	25017 samples	$[3.9115\mu s, 3.9127\mu s]$	25114 samples
$[4.0113\mu s, 4.0129\mu s]$	10653 samples	$[4.0115\mu s, 4.0131\mu s]$	10754 samples
Src: 3C905Tx — 3m CAT-5 — Dst: MII-NTI			
$[7.5558\mu s, 7.5588\mu s]$	25039 samples	$[7.5558\mu s, 7.5589\mu s]$	25038 samples
$[7.6565\mu s, 7.6592\mu s]$	25082 samples	$[7.6565\mu s, 7.6593\mu s]$	25039 samples
$[7.7572\mu s, 7.7596\mu s]$	24778 samples	$[7.7573\mu s, 7.7597\mu s]$	24803 samples
$[7.8576\mu s, 7.8600\mu s]$	25101 samples	$[7.8577\mu s, 7.8601\mu s]$	25120 samples
Src: 3C905Tx — 99m CAT-5 — Dst: MII-NTI			
$[8.0152\mu s, 8.0185\mu s]$	25079 samples	$[8.0153\mu s, 8.0186\mu s]$	25192 samples
$[8.1150\mu s, 8.1180\mu s]$	24745 samples	$[8.1154\mu s, 8.1184\mu s]$	24877 samples

continued on next page

continued from previous page

Full Duplex		Half Duplex	
$[8.2151\mu s, 8.2178\mu s]$	25256 samples	$[8.2152\mu s, 8.2179\mu s]$	24998 samples
$[8.3149\mu s, 8.3176\mu s]$	24920 samples	$[8.3153\mu s, 8.3177\mu s]$	24933 samples
Src: DFE530Tx — 3m CAT-5 — Dst: 3C905Tx			
$[1.7877\mu s, 1.7917\mu s]$	4437 samples	$[1.7872\mu s, 1.7912\mu s]$	4498 samples
$[1.8870\mu s, 1.8920\mu s]$	16833 samples	$[1.8870\mu s, 1.8911\mu s]$	16731 samples
$[1.9869\mu s, 1.9904\mu s]$	24927 samples	$[1.9864\mu s, 1.9899\mu s]$	25171 samples
$[2.0867\mu s, 2.0897\mu s]$	24961 samples	$[2.0863\mu s, 2.0888\mu s]$	24884 samples
$[2.1860\mu s, 2.1890\mu s]$	20614 samples	$[2.1857\mu s, 2.1887\mu s]$	20425 samples
$[2.2868\mu s, 2.2894\mu s]$	8228 samples	$[2.2860\mu s, 2.2890\mu s]$	8291 samples
Src: DFE530Tx — 99m CAT-5 — Dst: 3C905Tx			
$[2.2965\mu s, 2.3005\mu s]$	3039 samples	$[2.2962\mu s, 2.2997\mu s]$	2834 samples
$[2.3958\mu s, 2.4003\mu s]$	15549 samples	$[2.3960\mu s, 2.4000\mu s]$	15553 samples
$[2.4957\mu s, 2.4992\mu s]$	25256 samples	$[2.4958\mu s, 2.4988\mu s]$	25112 samples
$[2.5955\mu s, 2.5980\mu s]$	24939 samples	$[2.5956\mu s, 2.5981\mu s]$	25126 samples
$[2.6948\mu s, 2.6978\mu s]$	21607 samples	$[2.6948\mu s, 2.6973\mu s]$	22129 samples
$[2.7947\mu s, 2.7982\mu s]$	9610 samples	$[2.7951\mu s, 2.7976\mu s]$	9246 samples
Src: DFE530Tx — 3m CAT-5 — Dst: MII-NTI			
$[7.4330\mu s, 7.4345\mu s]$	24863 samples	$[7.4330\mu s, 7.4345\mu s]$	24997 samples
$[7.5335\mu s, 7.5350\mu s]$	24934 samples	$[7.5335\mu s, 7.5350\mu s]$	25020 samples
$[7.6339\mu s, 7.6355\mu s]$	24850 samples	$[7.6337\mu s, 7.6352\mu s]$	24949 samples
$[7.7341\mu s, 7.7356\mu s]$	25353 samples	$[7.7339\mu s, 7.7357\mu s]$	25034 samples
Src: DFE530Tx — 99m CAT-5 — Dst: MII-NTI			
$[7.9403\mu s, 7.9415\mu s]$	25007 samples	$[7.9401\mu s, 7.9416\mu s]$	24947 samples
$[8.0402\mu s, 8.0417\mu s]$	24584 samples	$[8.0398\mu s, 8.0419\mu s]$	24971 samples
$[8.1405\mu s, 8.1420\mu s]$	25227 samples	$[8.1404\mu s, 8.1419\mu s]$	24952 samples
$[8.2404\mu s, 8.2422\mu s]$	25182 samples	$[8.2401\mu s, 8.2423\mu s]$	25130 samples
Src: MII-NTI — 3m CAT-5 — Dst: 3C905Tx			
$[1.9123\mu s, 1.9168\mu s]$	4311 samples	$[1.9122\mu s, 1.9172\mu s]$	4314 samples
$[2.0121\mu s, 2.0166\mu s]$	16544 samples	$[2.0121\mu s, 2.0166\mu s]$	16658 samples
$[2.1115\mu s, 2.1160\mu s]$	24808 samples	$[2.1114\mu s, 2.1154\mu s]$	25308 samples
$[2.2113\mu s, 2.2143\mu s]$	25243 samples	$[2.2112\mu s, 2.2147\mu s]$	25007 samples
$[2.3112\mu s, 2.3137\mu s]$	20765 samples	$[2.3111\mu s, 2.3136\mu s]$	20431 samples
$[2.4115\mu s, 2.4140\mu s]$	8329 samples	$[2.4114\mu s, 2.4139\mu s]$	8282 samples
Src: MII-NTI — 99m CAT-5 — Dst: 3C905Tx			
$[2.4133\mu s, 2.4173\mu s]$	2377 samples	$[2.4132\mu s, 2.4172\mu s]$	2350 samples
$[2.5125\mu s, 2.5181\mu s]$	14803 samples	$[2.5130\mu s, 2.5175\mu s]$	14825 samples
$[2.6128\mu s, 2.6168\mu s]$	24925 samples	$[2.6127\mu s, 2.6167\mu s]$	25097 samples
$[2.7125\mu s, 2.7155\mu s]$	25029 samples	$[2.7124\mu s, 2.7154\mu s]$	24928 samples
$[2.8123\mu s, 2.8143\mu s]$	22600 samples	$[2.8121\mu s, 2.8141\mu s]$	22472 samples
$[2.9120\mu s, 2.9145\mu s]$	10266 samples	$[2.9119\mu s, 2.9144\mu s]$	10328 samples
Src: MII-NTI — 3m CAT-5 — Dst: AT2700Tx			
$[7.5515\mu s, 7.5539\mu s]$	24738 samples	$[7.5515\mu s, 7.5539\mu s]$	25139 samples

continued on next page

continued from previous page

Full Duplex		Half Duplex	
$[7.6516\mu s, 7.6543\mu s]$	25134 samples	$[7.6518\mu s, 7.6543\mu s]$	25004 samples
$[7.7520\mu s, 7.7544\mu s]$	25297 samples	$[7.7519\mu s, 7.7547\mu s]$	25063 samples
$[7.8521\mu s, 7.8548\mu s]$	24831 samples	$[7.8520\mu s, 7.8547\mu s]$	24794 samples
Src: MII-NTI — 99m CAT-5 — Dst: AT2700Tx			
$[8.0583\mu s, 8.0608\mu s]$	25179 samples	$[8.0586\mu s, 8.0610\mu s]$	25058 samples
$[8.1581\mu s, 8.1611\mu s]$	24974 samples	$[8.1586\mu s, 8.1613\mu s]$	25065 samples
$[8.2585\mu s, 8.2615\mu s]$	24962 samples	$[8.2588\mu s, 8.2616\mu s]$	24987 samples
$[8.3585\mu s, 8.3616\mu s]$	24885 samples	$[8.3588\mu s, 8.3615\mu s]$	24890 samples
Src: MII-NTI — 3m CAT-5 — Dst: DFE530Tx			
$[3.0045\mu s, 3.0061\mu s]$	15230 samples	$[3.0043\mu s, 3.0063\mu s]$	14900 samples
$[3.1042\mu s, 3.1062\mu s]$	25063 samples	$[3.1040\mu s, 3.1060\mu s]$	24904 samples
$[3.2043\mu s, 3.2063\mu s]$	24804 samples	$[3.2042\mu s, 3.2066\mu s]$	25292 samples
$[3.3044\mu s, 3.3064\mu s]$	24965 samples	$[3.3043\mu s, 3.3063\mu s]$	24956 samples
$[3.4045\mu s, 3.4065\mu s]$	9938 samples	$[3.4045\mu s, 3.4065\mu s]$	9948 samples
Src: MII-NTI — 99m CAT-5 — Dst: DFE530Tx			
$[3.6104\mu s, 3.6124\mu s]$	14260 samples	$[3.6109\mu s, 3.6121\mu s]$	14246 samples
$[3.7105\mu s, 3.7125\mu s]$	25009 samples	$[3.7109\mu s, 3.7121\mu s]$	25199 samples
$[3.8102\mu s, 3.8122\mu s]$	24978 samples	$[3.8108\mu s, 3.8120\mu s]$	25103 samples
$[3.9103\mu s, 3.9123\mu s]$	25064 samples	$[3.9108\mu s, 3.9120\mu s]$	24992 samples
$[4.0103\mu s, 4.0124\mu s]$	10689 samples	$[4.0108\mu s, 4.0124\mu s]$	10460 samples

Table 5.5: 10 Base-T delay measurements with Fast-Ethernet cards

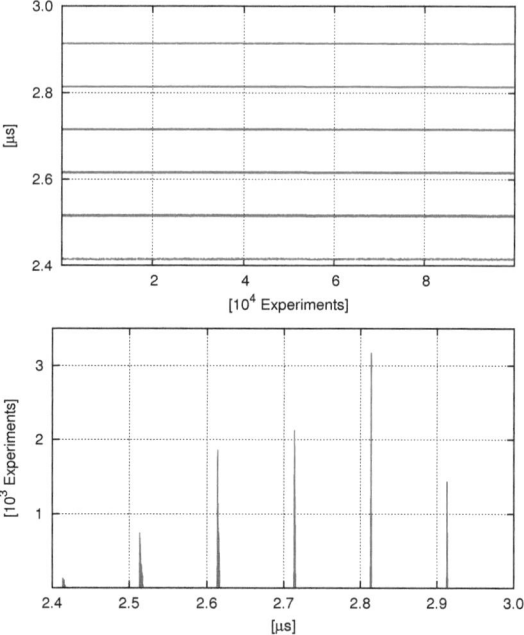

Figure 5.9: 10 Base-T Full Duplex Delay Measurements (top) and histogram (bottom) Sender: MII-NTI – 3m CAT-5 – Receiver: 3C905Tx

From the results we can deduce the following items for Fast-Ethernet cards in 10 Base-T mode:

- The measured delays are spread over four to six narrow intervals (1-4 ns wide) that are spaced 100 ns apart from each other. The number of intervals depend on the physical layer receiver.

- The number of measured samples are randomly distributed over these intervals and over time.

- The distributions of the values within the intervals are mostly Normal.

- The delays and delay variations are nearly the same for full-duplex and half-duplex mode.

- The employed CAT-5 cables add a delay proportional to the cable length. The delay variations due to the cables are negligible.

- The delays between the different configurations that are due to the physical layer devices vary by several μs.

The results clearly show that even given the best results of the delay variation obtained by NIC's equipped with the LXT970A physical layer device ($\sim 300ns$) one cannot reduce the remote clock reading error down to the ns-range. Hence, a precision of the clock synchronization clearly below the μs-range is not plausible with the given Fast-Ethernet cards for 10 Base-T mode. The reason for the delays being distributed across multiple intervals seem to stem from the fact that in 10 Base-T mode the receiver clock needs to be re-synchronized at the preamble of every received frame. In the given mode this synchronization is performed with a multiple (4x) of the oscillator frequency (2.5 MHz) supplied to the physical layer devices that directly relates to the 100 ns spacing of the measured intervals.

Next to Fast-Ethernet cards we surveyed also one type of a Gigabit-Ethernet card in 10 Base-T mode. These modern devices offer enhanced signal processing capabilities that may influence the expected results. Tab. 5.6 summarizes the results and Fig. 5.10 gives the histograms for full-duplex mode.

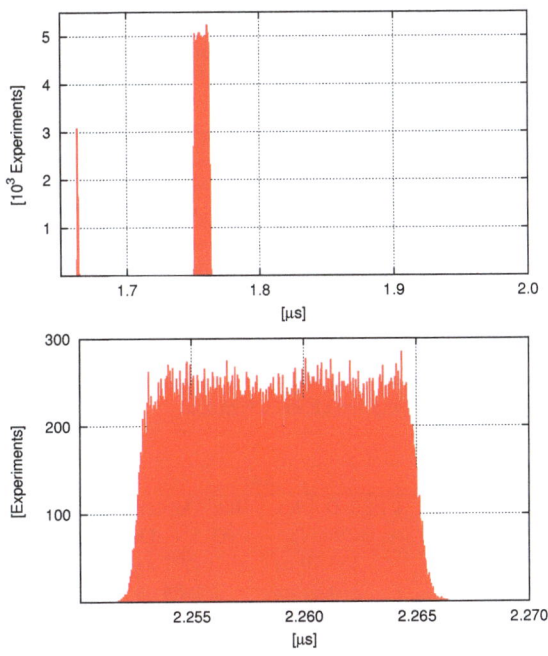

Figure 5.10: 10 Base-T Full Duplex histograms for Sender: DGE500T – 3m CAT-5 – Receiver: DGE500T (top) and Sender: DGE500T – 99m CAT-5 – Receiver: DGE500T (bottom)

Here most results are massed in one single interval (width $\sim 15ns$) and several outliers are found farther apart. In contrast to the results obtained using Fast-Ethernet cards the results within the main interval provide near Uniform distribution. Again the results for full-duplex and half-duplex look alike. This, however, is not the case when the delay varia-

Full Duplex		Half Duplex	
Src: DGE500T — 3m CAT-5 — Dst: DGE500T			
$[1.6503\mu s, 1.6518\mu s]$	13 samples	$[1.6507\mu s, 1.6518\mu s]$	97 samples
$[1.6618\mu s, 1.6643\mu s]$	5096 samples	$[1.6620\mu s, 1.6642\mu s]$	3991 samples
$[1.7502\mu s, 1.7636\mu s]$	94889 samples	$[1.7505\mu s, 1.7638\mu s]$	95912 samples
$[1.9619\mu s, 1.9625\mu s]$	15 samples		
Src: DGE500T — 99m CAT-5 — Dst: DGE500T			
		$[2.1539\mu s, 2.1556\mu s]$	12 samples
$[2.2517\mu s, 2.2663\mu s]$	10000 samples	$[2.2513\mu s, 2.2658\mu s]$	99988 samples

Table 5.6: 10 Base-T delay measurements with Gigabit-Ethernet cards

tion employing 3m CAT-5 cable are related to 99m CAT-5 cable. In the latter configuration almost all measurements are within one single interval.

Concluding, 10 Base-T mode of Gigabit-Ethernet devices seem more promising when compared to the Fast-Ethernet results, especially when one could employ some statistic filtering mechanisms to cut-off the outliers. In this case the delay variation is in the range of 15ns which promises a clock precision in the 100ns-range.

100 Base-Tx results: 100 Base-Tx Ethernet technology is most widely used in enterprize-wide office networks. In combination with switches these devices are usually operated in full-duplex mode. In fewer cases half-duplex mode using repeaters is still used to connect several computers within proximity, although more and more small desktop switches replace the repeaters here as well. Hence an investigation of 100 Base-Tx networks is of higher interest. Tab. 5.7 summarizes the results of the 100 Base-Tx delay measurements. As can be seen in the representative sample histogram in Fig. 5.11 the distribution of the results is Normal, therefore in Tab. 5.7 we provide typical statistical parameters. Next to minimum, maximum we extracted/calculated the median and the mean values, 25% and 75% quantiles, the standard deviation with its variation coefficient and the standard error of the mean to provide detailed characterization.

100 Base-Tx	3m		99m	
	Full-Duplex	Half-Duplex	Full-Duplex	Half-Duplex
Src: 3C905Tx — CAT-5 — Dst: 3C905Tx				
Minimum	269.79 ns	269.97 ns	772.2 ns	764.24 ns
Median	270.87 ns	270.88 ns	773.2 ns	765.31 ns
Maximum	271.72 ns	271.67 ns	774.28 ns	766.27 ns
Mean	270.85 ns	270.86 ns	773.21 ns	765.30 ns
Std. error of mean	0.78 ps	0.73 ps	0.88 ps	0.86 ps
25% quantile	270.68 ns	270.7 ns	773.01 ns	765.12 ns
75% quantile	271.03 ns	271.03 ns	773.4 ns	765.5 ns
Std. deviation	248 ps	232 ps	279 ps	271 ps
Variation coeff.	0.09%	0.09%	0.04%	0.04%
Src: 3C905Tx — CAT-5 — Dst: DFE530Tx[8]				

continued on next page

continued from previous page

100 Base-Tx	3m		99m	
	Full-Duplex	Half-Duplex	Full-Duplex	Half-Duplex
Minimum	302.4 ns	294.42 ns	784.62 ns	760.4 ns
Median	303.5 ns	295.49 ns	785.51 ns	761.52 ns
Maximum	304.86 ns	296.62 ns	786.44 ns	762.49 ns
Mean	303.5 ns	295.49 ns	785.52 ns	761.51 ns
Std. error of mean	0.92 ps	0.90 ps	0.75 ps	0.76 ps
25% quantile	303.3 ns	295.29 ns	785.35 ns	761.35 ns
75% quantile	303.7 ns	295.69 ns	785.67 ns	761.67 ns
Std. deviation	290 ps	286 ps	235 ps	238 ps
Variation coeff.	0.1%	0.1%	0.03%	0.03%
Src: 3C905Tx — CAT-5 — Dst: MII-NTI				
Minimum	311.78 ns	311.72 ns	825.18 ns	817.27 ns
Median	312.76 ns	312.75 ns	826.42 ns	818.43 ns
Maximum	313.71 ns	313.74 ns	827.91 ns	820.05 ns
Mean	312.76 ns	312.75 ns	826.47 ns	818.5 ns
Std. error of mean	0.77 ps	0.76 ps	1.15 ps	1.25 ps
25% quantile	312.6 ns	312.58 ns	826.21 ns	818.21 ns
75% quantile	312.93 ns	312.91 ns	826.69 ns	818.75 ns
Std. deviation	244 ps	240 ps	363 ps	397 ps
Variation coeff.	0.08%	0.08%	0.04%	0.05%
Src: DFE530Tx — CAT-5 — Dst: 3C905Tx				
Minimum	329.18 ns	313.18 ns	809.79 ns	799.69 ns
Median	330.43 ns	314.37 ns	810.95 ns	800.87 ns
Maximum	331.57 ns	315.47 ns	812.15 ns	802.09 ns
Mean	330.44 ns	314.37 ns	810.95 ns	800.87 ns
Std. error of mean	0.98 ps	0.90 ps	0.98 ps	0.97 ps
25% quantile	330.22 ns	314.17 ns	810.73 ns	800.65 ns
75% quantile	330.65 ns	314.56 ns	811.17 ns	801.09 ns
Std. deviation	310 ps	285 ps	310 ps	308 ps
Variation coeff.	0.09%	0.09%	0.04%	0.04%
Src: DFE530Tx — CAT-5 — Dst: MII-NTI				
Minimum	356.36 ns	388.37 ns	883.96 ns	867.97 ns
Median	357.28 ns	389.35 ns	884.99 ns	869.01 ns
Maximum	358.23 ns	390.27 ns	886.07 ns	869.95 ns
Mean	357.28 ns	389.34 ns	884.99 ns	869.01 ns
Std. error of mean	0.74 ps	0.73 ps	0.76 ps	0.74 ps
25% quantile	357.12 ns	389.19 ns	884.83 ns	868.85 ns
75% quantile	357.44 ns	389.5 ns	885.15 ns	869.17 ns
Std. deviation	233 ps	230 ps	240 ps	235 ps
Variation coeff.	0.07%	0.06%	0.03%	0.03%
Src: MII-NTI — CAT-5 — Dst: 3C905Tx				
Minimum	343.58 ns	343.69 ns	857.31 ns	857.46 ns

continued on next page

continued from previous page

100 Base-Tx	3m		99m	
	Full-Duplex	Half-Duplex	Full-Duplex	Half-Duplex
Median	344.78 ns	344.86 ns	858.23 ns	858.55 ns
Maximum	346.44 ns	346.51 ns	859.73 ns	859.77 ns
Mean	344.78 ns	344.87 ns	858.31 ns	858.55 ns
Std. error of mean	1.00 ps	1.01 ps	1.02 ps	1.00 ps
25% quantile	344.56 ns	344.64 ns	858.3 ns	858.33 ns
75% quantile	345 ns	345.09 ns	858.75 ns	858.77 ns
Std. deviation	318 ps	321 ps	322 ps	317 ps
Variation coeff.	0.09%	0.09%	0.04%	0.04%
Src: MII-NTI — CAT-5 — Dst: AT2700Tx				
Minimum	390.56 ns	382.46 ns	878.33 ns	902.31 ns
Median	391.65 ns	383.63 ns	879.5 ns	903.55 ns
Maximum	392.9 ns	384.8 ns	880.63 ns	904.78 ns
Mean	391.65 ns	383.63 ns	879.5 ns	903.55 ns
Std. error of mean	0.87 ps	0.91 ps	878.08 ns	0.91 ps
25% quantile	391.46 ns	383.43 ns	879.31 ns	903.35 ns
75% quantile	391.84 ns	383.82 ns	879.69 ns	903.75 ns
Std. deviation	275 ps	286 ps	278 ps	288 ps
Variation coeff.	0.07%	0.07%	0.03%	0.03%
Src: MII-NTI — CAT-5 — Dst: DFE530Tx[9]				
Minimum	371.19 ns	371.26 ns	767.89 ns	767.73 ns
Median	372.49 ns	372.42 ns	768.83 ns	768.79 ns
Maximum	373.82 ns	373.95 ns	769.86 ns	769.64 ns
Mean	372.49 ns	372.43 ns	768.83 ns	768.78 ns
Std. error of mean	1.05 ps	1.05 ps	0.76 ps	0.75 ps
25% quantile	372.26 ns	372.19 ns	768.66 ns	768.62 ns
75% quantile	372.72 ns	372.65 ns	768.99 ns	768.95 ns
Std. deviation	332 ps	333 ps	239 ps	236 ps
Variation coeff.	0.09%	0.09%	0.03%	0.03%
1000 Base-Tx	**3m**		**99m**	
Src: DGE500T — CAT-5 — Dst: DGE500T				
Minimum	618.16 ns	618.19 ns	$1.1158 \mu s$	$1.1156 \mu s$
Median	618.88 ns	618.89 ns	$1.1165 \mu s$	$1.1164 \mu s$
Maximum	620.23 ns	620.14 ns	$1.1179 \mu s$	$1.1178 \mu s$
Mean	618.89 ns	618.89 ns	$1.1165 \mu s$	$1.1164 \mu s$
Std. error of mean	0.55 ps	0.55 ps	0.56 ps	0.56 ps
25% quantile	618.77 ns	618.78 ns	$1.1164 \mu s$	$1.1163 \mu s$
75% quantile	618.99 ns	619 ns	$1.1167 \mu s$	$1.1165 \mu s$
Std. deviation	174 ps	174 ps	177 ps	176 ps
Variation coeff.	0.03%	0.03%	0.02%	0.02%

Table 5.7: 100 Base-Tx delay measurements with Fast- and Gigabit-Ethernet cards

From the given results one can deduce the following findings:

- The worst case delay variation is next to or below $3ns$ for all given configurations regardless of
  - the cable length
  - the operating mode (full-duplex/half-duplex)
  - wether the device employed is designed for Fast-Ethernet or Gigabit-Ethernet.

- The standard deviation is below $350ps$, hence the average delay variation is clearly below a $100MHz$ synchronizer error introduced due to different clocking domains of the receive and the transmit channel at the MII.

Hence, we can confirm by experiment that by employing the architecture proposed in Sec. 4.2 it is possible to improve existing clock synchronization techniques via existing wired communication systems by the order of several magnitudes down to the ns-range.

As a side-effect of these measurements we discovered that the DFE530Tx network interface card produced packet-loss by as much as 1.5% when operated over a 99m connection as receiver. Although this will have no influence on the clock reading error (these packets are simply dropped), it reduces throughput and clearly violates the Ethernet specification [39].

1000 Base-Tx results: Gigabit-Ethernet mainly employed in enterprize backbones will gain substantial significance with emerging multimedia techniques. Therefore, we conducted some experiments to analyze the delay variations caused by Gigabit physical layer devices. Tab. 5.8 summarizes our findings illustrated along with a representative sample plot in Fig. 5.12.

Full Duplex	Half Duplex
Src: DGE500T — 3m CAT-5 — Dst: DGE500T	
$[0.8751\mu s, 0.8836\mu s]$ 100000 samples	$[0.8745\mu s, 0.8836\mu s]$ 100000 samples
Src: DGE500T — 99m CAT-5 — Dst: DGE500T	
$[1.3686\mu s, 1.3772\mu s]$ 100000 samples	$[1.3602\mu s, 1.3692\mu s]$ 100000 samples

Table 5.8: 1000 Base-Tx delay measurements with Gigabit-Ethernet cards

The following findings can be deduced from these results:

- The distribution of the transmission delays for a 1000 Base-Tx point-to-point connection is Uniform with minimum and maximum laying $\sim 8ns$ apart.

- Again, the measured delay variation is independent from the employed cable length and operating mode.

From the presented results a precision in the *50ns-range* seem's feasible. Hence, these Gigabit devices provide reduced performance concerning the transmission delay variation and the derived remote clock reading error, especially because of the Uniform distribution.

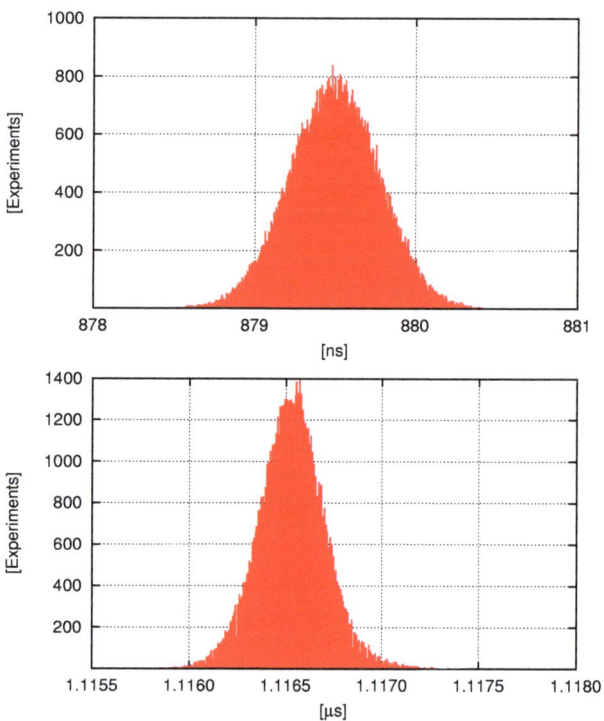

Figure 5.11: 100 Base-Tx Full Duplex histograms for Sender: MII-NTI – 99m CAT-5 – Receiver: AT2700Tx (top) and Sender: DGE500T – 99m CAT-5 – Receiver: DGE500T (bottom)

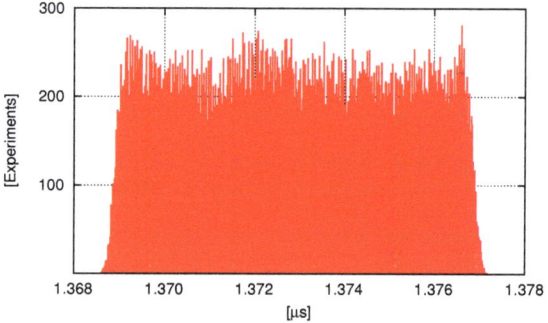

Figure 5.12: 1000 Base-Tx Full Duplex histogram Sender: DGE500T – 99m CAT-5 – Receiver: DGE500T

Linkloss Experiments: Up to now, all measurements were conducted under undisturbed office conditions. Since several application fields of distributed clock synchronization are in rough industrial environments we investigated the effect of loss and recovery of a link and how it could affect the results of the delay variations.

Electromagnetic noise can lead to a change of the voltages levels present at the cables during data transmission and that, in turn, can lead to bit flips and errors. However, it is the purpose of the 32-bit frame check sequence appended to every Ethernet frame to discover such situations. In case of an error detection frames are discarded (except for promiscuous mode) and an error is signaled to the driver software. This principal behavior doesn't deteriorate the achievable clock reading error as long as the characteristics of the physical layer devices and the cable stay unchanged. To that end we performed some simple experiments to discover the actual behavior.

First, we reused the setup of the previous experiments as illustrated in Fig. 5.8 and produced loss and recovery of the link by restarting auto-negotiation or by setting the media mode and speed anew several times while recording the delay variations before and after the link setting. The experiments for all 10 Base-T modes and configurations show the same results as without manipulation of the link status, hence no vial influence could be detected. In contrast, we encountered deviating results for 100 Base-Tx and 1000 Base-Tx settings. Fig. 5.13 shows the results of one typical configuration where the delay variation for 50.000 CSP's was recorded and loss of link and subsequent media negotiation were performed repetitively after the transmission of every 500th packet.

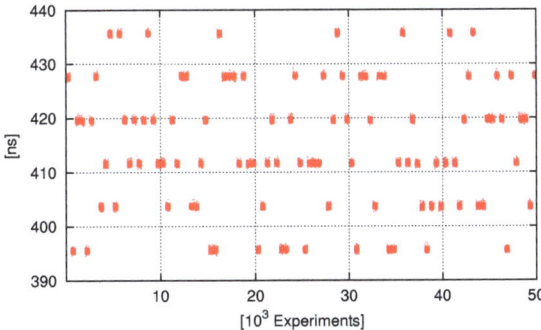

Figure 5.13: 100Base-Tx Full-Duplex Loss of Link Experiment; Sender: MII-NTI – 3m CAT-5 – Receiver: 3C905Tx

From the results we derive that for 100 Base-Tx the Normal distributions given by the direct connection experiments could be re-produced although the entire distributions are offset before and after the link change from each other by several ns. Results for all Fast-Ethernet cards show six distributions spaced about 8ns from each other meaning that the minimum and maximum measured delays were more than 40ns apart, see Fig. 5.13 for an example. This result would render tight clock synchronizations in the ns-range impossible using 100 Base-Tx based networks, hence the clock synchronization software must detect loss of link or restart of auto-negotiation and initiate subsequently a round-trip delay measurement to assess the actual delay. This effect is better when the Gigabit adapters are used; here only two Normal distributions are encountered in 100 Base-Tx mode and two Uniform distributions in 1000 Base-Tx mode either spaced by 8ns, see Fig. 5.14.

In order to understand this relocation of the distribution an inspection of the physical layer devices as done in Sec. 5.1.3 is appropriate. A digital PLL is used in the Fast-Ethernet PHY's and an analog PLL in the Gigabit-Ethernet PHY's to multiply the external 25MHz by a factor of at least 5 resulting in 125MHz, since every MII data nibble is serialized into 5 bit code groups. From the 8ns period of this 125MHz clock we suggest that the above effect is due to sampling and locking of the PLL's onto the received datastream.

In order to asses whether this effect can be triggered by the recovery of the link alone or also by external interference we conducted some further experiments. We used an EMC source to induce small voltage surges onto an unwinded twisted pair cable. Although these experiments produced controlled packet loss, we were not able to reproduce the effect following a loss of link.

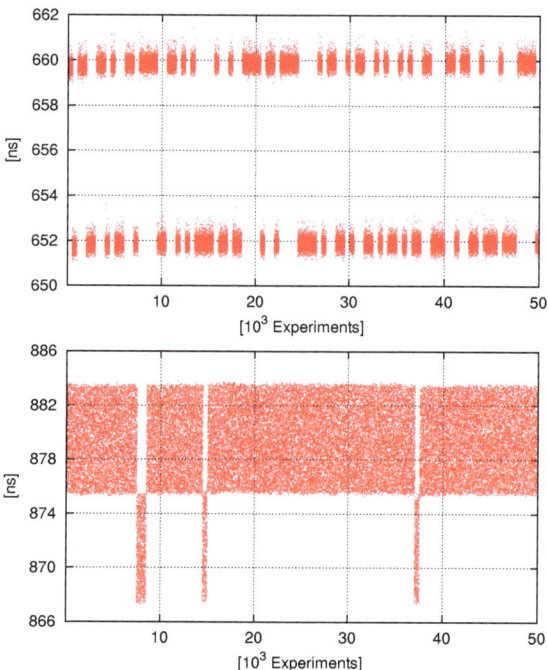

Figure 5.14: 100Base-Tx (top) and 1000 Base-Tx (bottom) Loss of Link Experiments with the DGE-500T adapters

5.3.2 Networked devices

The principle aim of the presented experiments was to assess the delay variation of clock synchronization packets due to the physical layer where two nodes are connected by a single cable. This is motivated by our architecture that is able to identify the variable delay of networked devices with our switch add-on, see Sec. 4.4, hence every end-to-end communications path is made up of a set of directly connected nodes. However, to substantiate the need for the additional logic for networked devices we collected some data on the delay variations of some COTS repeaters and switches. Tab. 5.9 lists the devices that were inspected under various different operating conditions.

Device	Short description
Repeaters	
3Com Hub 3C16704	4 ports: 10 Base-T
3Com Hub 3C16750B	8 ports: 10 Base-T and 100 Base-Tx (dualspeed)
Surecom 508T	8 ports: 10 Base-T and 1 port: AUI
Switches	
Cisco Catalyst 1900	24 ports: 10 Base-T and 2 ports: 100 Base-Tx (managable)
	Modes: Store & Forward, Fragment-Free Cut-Through
Surecom EP824-DX	24 ports: 10 Base-T and 100 Base-Tx (managable)
	Modes: Store & Forward

Table 5.9: Repeater and Switches, subjects of the evaluation

Repeater Experiments: A repeater, aka. as hub, synchronizes incoming data frames, amplifies and re-transmit them on all other ports. For this functionality the Ethernet specification grants maximum delays and delay variations, see Tab. 4.3. To that end, a repeater could add up to $80ns$ in 100 Base-Tx mode onto the transmission variation. We conducted several unloaded experiments with a setup as illustrated in Fig. 5.8 to get some real values for best-case conditions. In 10 Base-T mode the results shows three Uniform distributions for the 3Com 3C16704 and the Surecom 508T devices, see Tab. 5.10 for a listing of the acquired intervals.

Src: AT2700Tx — 2m CAT-5 — Hub — 2m CAT-5 — Dst: MII-NTI			
3Com 3c16704		**Surecom 508T**	
$[8.6200\mu s, 8.6755\mu s]$	13227 samples	$[9.1449\mu s, 9.1757\mu s]$	7491 samples
$[8.7210\mu s, 9.0245\mu s]$	74993 samples	$[9.2456\mu s, 9.5477\mu s]$	75043 samples
$[9.0760\mu s, 9.1250\mu s]$	11780 samples	$[9.5764\mu s, 9.6484\mu s]$	17466 samples

Table 5.10: 10 Base-T delay measurements with Hubs

The difference between the maximum and the minimum transmission delays in the range of $500ns$ for unloaded configurations and the Uniform distribution clearly forestall clock synchronization precisions below the μs-range.

The results for the dualspeed hub 3C16750B show five Uniform distributions in 10 Base-T mode and one Uniform distribution in 100 Base-Tx mode. The given repeater can operate either link with dissimilar speeds; experiments using this configuration show Trapezoid distributions. From Tab. 5.11 one can see that the differences between the maximum and minimum transmission delays exceeds $700ns$ for 10 Base-T and $40ns$ for 100 Base-Tx forestalling clock precisions in the range of below $160ns$ as well.

The transmission delay variation added by the presented repeaters are within the IEEE specification [39], but the given values render clock precision in the ns-range impossible. Since repeaters are frequently replaced by switches we didn't propose specific architectural concepts to enhance them with support for tight clock synchronization; however, the

Src: AT2700Tx — 2m CAT-5 — Hub — 2m CAT-5 — Dst: MII-NTI		
both links 10 Base-T	$[8.3206\mu s, 8.3777\mu s]$	13317 samples
	$[8.4200\mu s, 8.4779\mu s]$	13022 samples
	$[8.5195\mu s, 8.5781\mu s]$	13150 samples
	$[8.6204\mu s, 8.7255\mu s]$	24928 samples
	$[8.7728\mu s, 8.8257\mu s]$	11769 samples
	$[8.8723\mu s, 8.9259\mu s]$	11830 samples
	$[8.9725\mu s, 9.0261\mu s]$	11984 samples
both links 100 Base-Tx	$[738.8ns, 782.3ns]$	100000 samples
10 Base-T and 100 Base-Tx	$[15.4853\mu s, 16.0712\mu s]$	100000 samples

Table 5.11: Delay measurements with Hub 3Com 3c16750B

concepts proposed for switches in Sec. 4.4 could be re-used for repeaters in a similar way.

Switches: As presented in Sec. 4.1 switches are at the core of every modern enterprize network. Therefore, we proposed some concepts to enhance COTS switches with support for tight clock synchronization, see Sec. 4.4. To substantiate this proposal we conducted some experiments using two different switches listed in Tab. 5.9. In particular, we investigated the influence of cut-through and store & forward techniques on the transmission delay variation of 50000 CSP's under unloaded and lightly loaded conditions in full-duplex mode. Again, the same setup as illustrated in Fig. 5.8 was used, although for the loaded scenario one additional PC was used for traffic generation directed to the receiving node of CSP transmissions.

Cut-Through Mode: Using a serial console interface, the Cisco switch was first programmed to cut-through operating mode. For the 10 Base-T experiments both nodes encompassing CSP traffic and the load generator were connected to one port group. The distribution of the results for the unloaded configuration shows a triangular shape with about 99.7% of the results and a somewhat longer tail where the remaining values are uniformly distributed. For the loaded experiments we used our traffic generation program to generate some load directed to the receiving port of CSP transmissions. The pattern of the load consisted of packets similar in size and repetition frequency to the CSP's. Again, about 98.7% of the packets were cumulated within a distribution showing triangular shape. However, when compared to the unloaded configuration the tail of the remaining uniformly distributed values is several orders of magnitude longer. In detail, the delay of the slowest packet doubles, see Tab. 5.12.

For the 100 Base-Tx experiments we connected the links encompassing CSP's to the only two 100 Base-Tx ports and used one 10 Base-T port for load generation in the same way as for the 10 Base-T experiments. The distribution of the results shows the same triangular shape and a similar tail. Fig. 5.15 opposes the conducted results for the unloaded and loaded configurations respectively, and Fig. 5.16 illustrates the distribution for the unloaded 100 Base-Tx configuration.

Store & Forward Mode: The majority of switches is built with store & forward techniques that enable packet filtering, prioritizing etc. To assess and quantify the impact of these techniques onto the CSP transmission variation we conducted some experiments with a Cisco Catalyst 1900 switch configured to store & forward mode and a Surecom EP824-DX switch. The experimental setup for the Cisco Catalyst 1900 switch was the same as for cut-through mode; for the Surecom device all three ports –those encompass-

unloaded	10 Base-T		100 Base-Tx	
distribution	$[54.28\mu s, 55.02\mu s]$	49839 samples	$[7.08\mu s, 7.17\mu s]$	49155 samples
tail	$[55.02\mu s, 55.41\mu s]$	161 samples	$[7.17\mu s, 7.31\mu s]$	845 samples
loaded	**10 Base-T**		**100 Base-Tx**	
distribution	$[54.27\mu s, 55.39\mu s]$	49353 samples	$[7.08\mu s, 7.31\mu s]$	49893 samples
tail	$[55.39\mu s, 121.58\mu s]$	647 samples	$[7.31\mu s, 17.39\mu s]$	107 samples

Table 5.12: Fragment free cut-through results for the Cisco Catalyst 1900

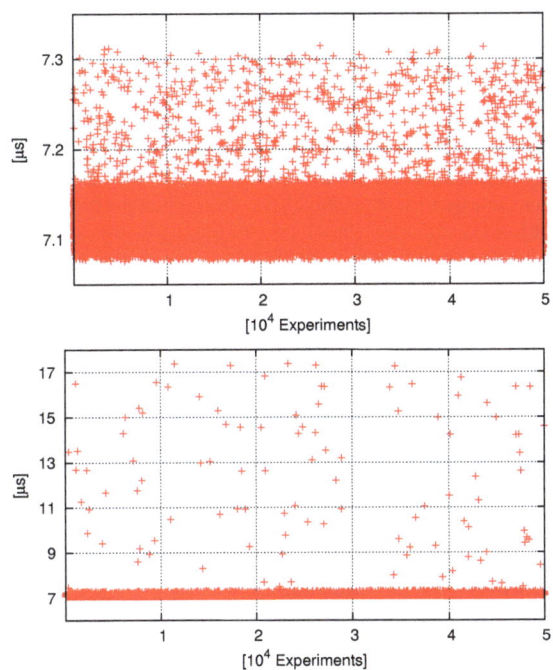

Figure 5.15: Results for 100 Base-Tx unloaded (top) and loaded (bottom) cut-through experiments

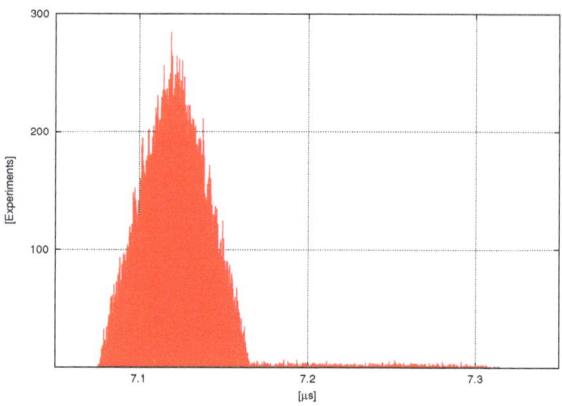

Figure 5.16: Result distribution for 100 Base-Tx unloaded cut-through experiments

ing CSP traffic and the load generator– were connected to one 100 Base-Tx port group. In either configuration we recorded the delay variation of 20000 CSP's.

The results for the Cisco switch show the same form of a triangular shaped distribution with a tail for unloaded, loaded, 10 Base-T and 100 Base-Tx configurations, see Tab. 5.13. The results for the Surecom switch show a Uniform distribution for the unloaded configuration. For the experiments in which the destination port of CSP's was loaded, we got a similar Uniform distribution for 99.1% of all transmission delays. The remaining values were equally distributed along a wide interval accounting for packets that were delayed when the receiving port was congested. Fig. 5.17 illustrates the results for the Surecom switch in 100 Base-Tx mode.

Cisco Catalyst 1900				
10 Base-T	unloaded		loaded	
distribution	$[54.17\mu s, 54.87\mu s]$	19828 samples	$[54.14\mu s, 55.17\mu s]$	19676 samples
tail	$[54.87\mu s, 55.25\mu s]$	172 samples	$[55.17\mu s, 156.53\mu s]$	324 samples
100 Base-Tx	unloaded		loaded	
distribution	$[7.13\mu s, 7.22\mu s]$	19643 samples	$[7.13\mu s, 7.36\mu s]$	19960 samples
tail	$[7.22\mu s, 7.36\mu s]$	357 samples	$[7.36\mu s, 16.37\mu s]$	40 samples
Surecom EP824-DX				
100 Base-Tx	unloaded		loaded	
distribution	$[6.79\mu s, 7.31\mu s]$	20000 samples	$[6.82\mu s, 7.29\mu s]$	19821 samples
tail			$[7.29\mu s, 49.22\mu s]$	179 samples

Table 5.13: Store & Forward results

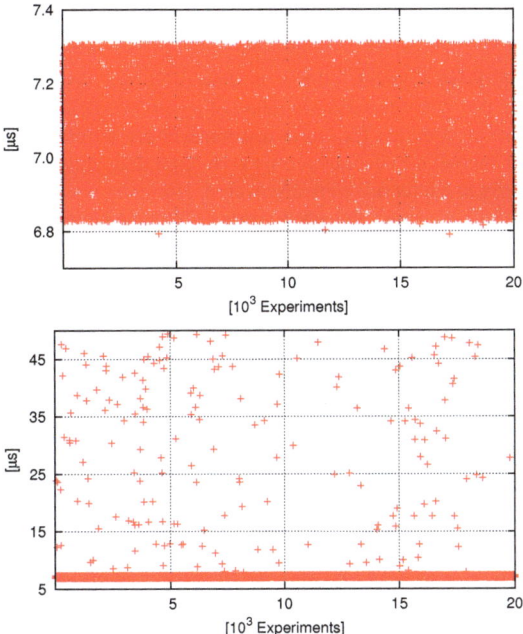

Figure 5.17: Results for 100 Base-Tx unloaded (top) and loaded (bottom) store & forward experiments

The results show that in unloaded conditions switches add several 100ns onto the delay variation of packet transmission. With a light load however the added delay variation changes to $10\mu s$ and more. This, in turn, directly relates to the remote clock reading error ε that impairs the achievable clock precision with at least a factor of 4. This clearly substantiates the quite apparent need for a switch architecture as proposed in Sec. 4.4.

5.4 Summary

This section investigated the influence of COTS Ethernet devices on the delay variation of clock synchronization packets between the sending and receiving timestamps at the media independent interface which resembles the remote clock reading error ε. Following an inspection and modeling of the physical layer devices and the cabling, we presented results of a set of extensive experiments with several COTS devices. These results lead to the following implications:

- Fast Ethernet PHY's in 10 Base-T mode give delays equally distributed over 4-6 intervals spaced about 100ns apart from each other. The delay variation due to these devices exceeds $300ns$, thus rendering clock precision below $1\mu s$ impossible.

- The Gigabit Ethernet PHY's in 10 Base-T mode produce Uniform distributions with most values accumulated within one interval of $15ns$ width. Few values, however, are apart by some 100*ns*. When these outliers can be filtered by the clock synchronization algorithm, a clock precision in the range of 100*ns* seems possible.

- In 100 Base-Tx mode all results show a Normal distribution with a difference between the maximum and minimum of about 2*ns*. Following a loss of link or a restart of auto-negotiation the entire distributions are separated by a multiple of 8*ns*. For Fast Ethernet we encountered six such separated intervals whereas for Gigabit Ethernet only two intervals were discovered. However, under normal operating conditions we didn't encounter this phenomenon. To that end, we reason that if we employ 100 Base-Tx devices we can keep the remote clock reading error in the ns-range if the loss of link or restart of auto-negotiation is signaled to the clock synchronization software.

- In 1000 Base-Tx mode the results have Uniform distribution with maximum and minimum delay laying $\sim$ 8*ns* apart. Following a loss of link or a restart of auto-negotiation we encountered the same dislocation phenomenon as for the 100 Base-Tx mode; here the entire distribution is dislocated by 8*ns*. Although the distribution is less promising than for 100 Base-Tx modes one can still bound the clock reading error by about 8*ns* as long as loss of link and restart of auto-negotiation is detected and handled by the clock synchronization software.

Furthermore, several experiments using COTS repeaters and switches were conducted to uncover some values these devices would add onto the transmission delay variation of clock synchronization packets. In unloaded configurations these values are more than 40*ns* and 100*ns* for the investigated repeaters and switches respectively. Experiments with a realistic light load show that these values increase by an order of several magnitudes.

Summarizing the results of the presented experiments clearly substantiate the need for the proposed hardware support for network interface cards and networked devices proposed in Chap. 4. Furthermore, from the results one can reason that if the proposed architectures are implemented, distributed clock synchronization in the 100*ns*-range is feasible.

Example

Finally, we present a numerical example with decently realistic numbers to underpin the presented results. For this example we use the formula for the worst-case precision given by Equ. 2.3 with the constants $c_1 = 4$, $c_2 = 4$, $c_3 = 3$, $c_4 = 11$ and $c_5 = 1$ (the result of the worst-case precision analysis for the Orthogonal Precision algorithm). We assume a set of nodes interconnected with one switch (n=1) and

- a worst-case jitter of one physical layer device $\varepsilon_{dmax} = 3ns$,
- a timestamp granularity $G_{ts} = 2^{-32}$,
- a worst-case transit delay through one switch $\delta_{swmax} = 1ms$,
- an oscillator drift $\rho_{sw} = 10^{-7} s/s$ (the same at every switch),
- a re-synchronization period of $P = 30s$,
- a clock setting granularity of $G_s = 2^{-64}$,

- a rate adjustment uncertainty and clock granularity of $u = G = 1/f_s = 10 ns$.

Using Equ. 4.10 and 2.3 Tab. 5.14 presents the achievable worst-case precision for different clock/sampling frequencies (the same at every node) and oscillator drifts (at the NICs).

$f_{osc} = f_s$	ρ_{nic}	4ε	$4P\rho_{nic}$	$3G$	$11u$	π
100 MHz	10 ppb	43 ns	1.2 μs	30 ns	110 ns	1.383 μs
	1 ppb		120 ns			303 ns
	0.1 ppb		12 ns			195 ns
200 MHz	10 ppb	28 ns	1.2 μs	15 ns	55 ns	1.298 μs
	1 ppb		120 ns			218 ns
	0.1 ppb		12 ns			110 ns
500 MHz	10 ppb	19 ns	1.2 μs	6 ns	22 ns	1.247 μs
	1 ppb		120 ns			167 ns
	0.1 ppb		12 ns			59 ns
1 GHz	10 ppb	16 ns	1.2 μs	3 ns	11 ns	1.23 μs
	11 ppb		120 ns			150 ns
	0.1 ppb		12 ns			42 ns

Table 5.14: Worst-case precision analysis for a typical network

The presented results clearly show that the oscillator parameters at the NICs dominate the achievable worst-case precision. Hence, either an excellent OCXO or a TCXO + rate clock synchronization algorithm must be employed to minimize the effect of the clock drift to an order of $10^{-9} s/s$. Next to the clock drift one should optimize the clock and sampling frequency at every node to reduce the effects of the clock reading error, the granularity and the rate adjustment uncertainty.

In practice, however, the average-case precision will be much lower. In fact, a precision in the several *ns*-range is plausible even for lower sampling frequencies and oscillators with moderate drift.

Chapter 6
Conclusion and Future Work

An accurate clock synchronization service is a fundamental pre-requisite for a distributed real time system. The two parameters clock precision π and clock accuracy α are of primary interest to characterize the performance of this service. Herein clock precision denotes the deviation of the clock states between any two nodes within the system, and clock accuracy specifies the clock state deviation between any node within the system and an external reference time. To that end many algorithms have been presented in scientific literature to optimize these parameters. Following the taxonomy presented in Chap. 2 the achievable worst case precision mainly depends on the remote clock reading error. When ε can be reduced below the μs-range, the clock drift ρ, the re-synchronization period P, the clock granularity G and the rate adjustment uncertainty u need to be considered as well.

Recent popular systems that rely on such a clock synchronization service with a precision in the μs-range are the time-triggered protocol TTP and FlexRay that are employed in future airborne and automotive systems. For networked measurement and control systems the new IEEE-1588 standard proposes a master-slave based protocol termed precision time protocol (PTP). Following a short analysis of these system we propose a new network interface architecture and support for networked devices tailored for twisted-pair based Ethernet systems. The main contributions of this architecture are

- transparent media-independent interface based timestamping,

- on-the-fly measurement of packet transmission delays through switches and

- the need for a high resolution, high frequency clock to minimize the synchronization error due to the different clocking domains of receive and transmit path.

The presented architecture improves the achievable worst-case precision of existing software-based approaches (e.g. NTP) by several orders of magnitude and will outperform the new IEEE-1588 standard as well, which disregards several aspects revealed in scientific literature. Our proposal is validated by an experimental evaluation of the remote clock reading error achievable with COTS Ethernet devices. The results revealed the following facts:

- 10 Base-T based networks produce delay variations equally distributed over several intervals spaced by $\sim 100 ns$.

- 100 Base-Tx based networks show a Normal distribution with $\varepsilon \sim 2ns$.

- 1000 Base-Tx networks show a Uniform distribution with $\varepsilon \sim 8ns$.

- The contribution of repeaters and switches onto ε exceeds several ten and hundred *ns* in an unloaded configuration. These values degrade to the *μs*-range when a load is present.

From the given results a worst-case clock precision in the 100*ns*-range seems feasible, when the proposed architecture is employed. To that end an integration of the proposed architecture with COTS components is required. This non-trivial engineering task is currently the main focus of the FIT-IT funded PSynUTC project. Following an industrial implementation and proper system evaluation several applications can be built that benefit from such an accurate clock synchronization service.

Next to an industrial implementation a detailed inspection on the susceptibility to electromagnetic emissions is required to judge whether the dislocation of the delay results following the loss and recovery of the link could occur in a rough environment as well.

Bibliography

[1] ANCEAUME, E., AND PUAUT, I. Performance Evaluation of Clock Synchronization Algorithms. Tech. Rep. PI 1103, Institute de Rechereche en Informatique et Systemes Aleatoires, July 1997.

[2] ARVIND, K. Probabilistic clock synchronization in distributed systems. *IEEE Transactions on Parallel and Distributed Systems 5*, 5 (May 1994), 474–487.

[3] AZEVEDO, M., AND BLOUGH, D. Fault-Tolerant Clock Synchronization for Distributed Systems with High Message Delay Variation. In *IEEE Workshop on Fault-Tolerant Parallel and Distributed Systems* (College Station, Texas, June 13-14 1994).

[4] AZEVEDO, M., AND BLOUGH, D. Multistep Interactive Convergence: An Efficient Approach to the Fault-Tolerant Clock Synchronization of Large Multicomputers. *IEEE Transactions on Parallel and Distributed Systems 9*, 12 (December 1998), 1195–1212.

[5] BALLATO, A., AND VIG, J. Static and dynamic frequency-temperature behavior of singly and doubly rotated, oven-controlled quartz resonators. In *Proceedings 32nd Symp. on Frequency Control* (1978), vol. 1, pp. 180–188.

[6] BARNES, J., CHI, A., CUTLER, L., HEALEY, D., LEESON, D., MCGUNIGAL, T., MULLEN, J., SMITH, W., SYNDOR, R., VESSOT, R., AND WINKLER, G. Charcterization of frequency stability. *IEEE Transactions on Instrument Measurement IM-20*, 2 (May 1971), 105–120.

[7] BERGER, J. *Fehlerortung in Mittelspannungsnetzen mit Abzweigleitungen*. Dissertation, University of Stuttgart, 1995. (in German).

[8] BO, Z., WELLER, G., JIANG, F., AND YANG, Q. Application of GPS based fault location scheme for distribution system. In *Proceedings on Power System Technology* (1998), vol. 1, pp. 53–57.

[9] BRENDEL, R. Infuence of magnetic field on quartz crystal oscillators. In *Proc. 43rd Ann. Symp. Frequency Control* (1989), pp. 268–274.

[10] CHERUBINI, G., ÖLCER, S., UNGERBOECK, G., CREIGH, J., AND RAO, S. 100BASE-T2: A New Standard for 100 Mb/s Ethernet Transmission over Voice-Grade Cables. *IEEE Communications Magazine* (November 1997), 115–122.

[11] CHIEN, G. *Low-Noise Local Oscillator Design Techniques using a DLL-based Frequency Multiplier for Wireless Applications*. PhD thesis, University of California at Berkeley, 2000.

[12] CHOI, B., PARK, K., AND KIM, M. An Improved Hardware Implementation of the Fault-Tolerant Clock Synchronization Algorithm for Large Multiprocessor System. *IEEE Transaction on Computers 39* (1990), 404–407.

[13] COMMISSION, F. C. FCC requires wireless carriers to forward all 911 calls, December 1997.

[14] COUVET, D., FLORIN, G., AND NATKIN, S. A statistical clock synchronization algorithm for anisotropic networks. In *10th Symposium on Reliable Distributed Systems* (1991), pp. 42–51.

[15] CRISTIAN, F. Probabilistic Clock Synchronization. *Distributed Computing 3*, 3 (1989), 146–158.

[16] CRISTIAN, F., AGHILI, H., AND STRONG, R. Clock synchronization in the presence of omission and performance failures, and processor joins. *In Proc. of 16th International Symposium on Fault-Tolerant Computing Systems* (July 1986).

[17] CRISTIAN, F., AND FETZER, C. Fault-tolerant internal clock synchronization. In *Proceedings of the Thirteenth Symposium on Reliable Distributed Systems* (Dana Point, Ca., Oct 1994), pp. 22–31.

[18] CRISTIAN, F., AND FETZER, C. Probabilistic internal clock synchronization. In *13th Symposium on Reliable Distributed Systems* (1994), pp. 22–31.

[19] DALLY, W., AND J.W.POULTON. *Digital Systems Engineering*. Cambridge University Press, 1998.

[20] DEWE, M., SANKAR, S., AND ARILLAGA, J. The application of satellite time references to HVDC fault location. *IEEE Transactions on Power Delivery 8* (1993), 1295–1302.

[21] DUDA, A., HARRUS, G., HADDAD, Y., AND BERNARD, G. Estimating global time in distributed systems. In *Conference on Distributed Computing Systems* (Berlin, 1987).

[22] EIDSON, J., FISCHER, M., AND WHITE, J. IEEE-1588 Standard for a Precision Clock Synchronization Protocol for Networked Measurement and Control Systems. In *Proceedings of the 34th Precise Time and Time Interval (PTTI) Systems and Applications Meeting* (Reston, Virginia, USA, Dec 2002).

[23] FETZER, C., AND CRISTIAN, F. Lower Bounds for Function Based Clock Synchronization. In *Proceedings 14th ACM Symposium on Principles of Distributed Computing* (Ottawa, CA, August 1995).

[24] FETZER, C., AND CRISTIAN, F. An Optimal Internal Clock Synchronization Algorithm. In *Proceedings 10th Annual IEEE Conference on Computer Assurance* (Gaithersburg, MD, June 1995).

[25] FETZER, C., AND CRISTIAN, F. Integrating External and Internal Clock Synchronization. *J. Real-Time Systems 12*, 2 (March 1997), 123–172.

[26] GAI, P. Cable Fault Location by Impulse Current Method. In *Proceedings of IEE* (Apr 1975), vol. 122, pp. 403–408.

[27] GUSELLA, R., AND ZATTI, S. The accuracy of the clock synchronization achieved by TEMPO in berkeley UNIX 4.3BSD. *IEEE Transactions on Software Engineering 15*, 7 (July 1989), 847–853.

[28] HAAG, H.-J. Quantensprung in der Messtechnik zur Bewertung der Netzqualität. In *Elektrotechnik und Informationstechnik e&i* (Wien, 2000), pp. 645–652.

[29] HÖCHTL, D., AND SCHMID, U. Long-Term Evaluation of GPS Timing Receiver Failures. In *Proceedings of the 29th IEEE Precise Time and Time Interval Systems and Application Meeting (PTTI'97)* (Long Beach, California, Dec 1997), pp. 165–180.

[30] HORAUER, M. Hardware support for clock synchronization in distributed systems. In *Supplement of the 2001 International Conference on Dependable Systems and Networks* (Göteborg, Sweden, July 2001), pp. A10–A13.

[31] HORAUER, M., AND HÖLLER, R. Integration of high accurate clock synchronization into ethernet-based distributed systems. In *International Conference on Advances in Infrastructure for e-Business, e-Education, e-Science, and e-Medicine on the Internet, SSGRR 2002* (L'Aquila, Italy, January 2002).

[32] HORAUER, M., KERÖ, N., AND SCHMID, U. A network interface for highly accurate clock synchronization. In *Proceedings AUSTROCHIP'00* (Graz, Austria, Oct 2000).

[33] HORAUER, M., LOY, D., AND SCHMID, U. NTI Functional and Architectural Specification. Tech. Rep. 183/1-69, TUAuto, December 1996.

[34] HORAUER, M., SCHMID, U., AND SCHOSSMAIER, K. NTI: A Network Time Interface M-Module for High-Accuracy Clock Synchronization. In *Proceedings 6th International Workshop on Parallel and Distributed Real-Time Systems (WP-DRTS'98)* (Orlando, Florida, March 30 – April 3 1998), pp. 1067–1076.

[35] HORAUER, M., SCHMID, U., SCHOSSMAIER, K., HÖLLER, R., AND KERÖ, N. Psynutc - evaluation of a high precision time synchronization prototype system for ethernet lans. In *Proceedings of the 34th IEEE Precise Time and Time Interval Systems and Application Meeting (PTTI'02)* (Reston, Virginia, USA, December 2002).

[36] HOWE, D., ALLAN, D., AND BARNES, J. Properties of signal sources and measurement methods. In *Proceedings of the 35th Annual Frequency Control Symposium* (1981).

[37] HOYME, K., AND DRISCOLL, K. Safebus. In *Proceedings of the 11th AIAA/IEEE Digital Avionics Systems Conference* (Seattle, Washington, USA, October 1992), pp. 68–73.

[38] HOYME, K., AND DRISCOLL, K. Safebus. *IEEE Aerospace and Electronic Systems Magazine 8*, 3 (March 1993), 34–39.

[39] IEEE. Carrier sense multiple access with collision detection (csma/cd) access method and physical layer specifications - ieee std 802.3, 2000 edition. Tech. rep., IEEE Computer Society, 2000.

[40] IEEE. 1588 IEEE Standard for a Precision Clock Synchronization Protocol for Networked Measurement and Control Systems. Tech. rep., IEEE Instrumentation and Measurement Society, 2002.

[41] INC., A. R. Arinc specification 659: Backplane data bus. Tech. rep., Airlines Electronic Engineering Committee, December 1993.

[42] INOUE, N., TSUNEKAGE, T., AND SAKAI, S. On-line fault location system for 66kV underground cables with fast O/E and fast A/D technique. *IEEE Transactions on Power Delivery 9*, 1 (January 1994), 579–584.

[43] INTEMATIONAL RADIO CONSULTATIVE COMMITTEE (CCIR) - STANDARD FREQUENCIES AND TIME SIGNALS (STUDY GROUP 7). *Recommendation No. 686, Glossary*, ccir 17th plenary assembly ed. Geneva, Switzerland, 1990.

[44] JEZEQUEL, M. Building a global time on parallel machines. Tech. Rep. Rapport de recherche - 513, INRIA, February 1990.

[45] KAMPICHLER, W. *Measurement of Voice Readiness in Local Area Communication Systems*. PhD thesis, University of Technology Vienna, Department of Computer Technology, 2002.

[46] KANG, G., AND RAMANATHAN, P. Clock Synchronization of a Large Multiprocessor System in the Presence of Malicious Faults. *IEEE Transactions on Computers 36*, 1 (January 1987), 2–12.

[47] KERÖ, N., SCHMID, U., AND HORAUER, M. Verfahren für die Synchronisation von Computeruhren in Netzwerken. Tech. Rep. 183/1-105, Department of Automation, TU Vienna, March 2000. Patent: AT005327U1.

[48] KIECKHAFER, R., WALTER, C., FINN, A., AND THAMBIDURAI, P. The MAFT Architecture for Distributed Fault Tolerance. *IEEE Transactions on Computers 37*, 4 (1988), 398–405.

[49] KOPETZ, H., DAMM, A., KOZA, C., MULAZZANI, M., SCHWABL, W., SENFT, C., AND ZAINLINGER, R. Distributed fault-tolerant real-time systems: The mars approach. *IEEE Micro 9*, 1 (February 1989), 25–50.

[50] KOPETZ, H., AND GRÜNSTEIDL, G. TTP - A Protocol for Fault-Tolerant Real-Time Systems. In *The Twenty-Third International Symposium on Fault-Tolerant Computing FTCS-23* (Aug. 1993), pp. 524–533.

[51] KOPETZ, H., AND GRÜNSTEIDL, G. TTP - a protocol for fault-tolerant real-time systems. *IEEE Computer 27*, 1 (January 1994), 14–23.

[52] KOPETZ, H., HEXEL, R., KRÜGER, A., MILLINGER, D., AND SCHEDL, A. A Synchronization Strategy for a TTP/C Controller. In *In Application of Multiplexing Technology, SAE International* (Feb. 1996), pp. 19–27.

[53] KOPETZ, H., AND OCHSENREITER, W. Clock Synchronization in Distributed Real-Time Systems. *IEEE Transactions on Computers C-06*, 8 (1987), 833–839.

[54] KUNKEL, J. *Fehlerortung im Hochspannungsnetz.* Dissertation, University of Stuttgart, 1990. (in German).

[55] LAMPORT, L. Synchronizing Time Servers. Technical Report 18, Digital System Research Center, 1980.

[56] LAMPORT, L., AND MELLIAR-SMITH, P. Synchronizing clocks in the presence of faults. *Journal of the ACM 32* (July 1985), 52–78.

[57] LAMPORT, L., SHOSTAK, R., AND PEASE, M. The byzantine generals problem. *ACM Transactions on Programming Languages and Systems 4* (July 1982), 382–401.

[58] LEVINE, J. Time synchronization using the Internet. *IEEE Transactions on Ultrasonics, Ferroelectrics and Frequency Control 45* (March 1998), 450–460.

[59] LISKOV, B. Practical uses of syhchronized clocks in distributed systems. *Distributed Computing 1* (1993), 211–211.

[60] LIU, C., JUNG, J., HEYDT, G., VITTAL, V., AND PHADKE, A. The Strategic Power Infrastructure Defense (SPID) System. *IEEE Control Systems Magazine* (Aug 2000), 25–52.

[61] LOY, D. *GPS-Linked High Accuracy NTP Time Processor for Distributed Fault-Tolerant Real-Time Systems.* Dissertation, Vienna University of Technology, Faculty of Electrical Engineering, Apr 1996.

[62] LUNDELIUS-WELCH, J., AND LYNCH, N. An Upper and Lower Bound for Clock Synchronization. *Information and Control 62* (1984), 190–209.

[63] LUNDELIUS-WELCH, J., AND LYNCH, N. A New Fault-Tolerant Algorithm for Clock Synchronization. *Information and Computation 77*, 1 (1986), 1–96.

[64] MALONEY, C. Locating Cable Faults. *IEEE Transactions on Industry Applications IA-9*, 4 (July/August 1973), 380–394.

[65] MARZULLO, K. *Maintaining the Time in a Distributed System: An Example of a Loosely-Coupled Distributed Service.* PhD dissertation, Stanford University, Department of Electrical Engineering, Feb 1984.

[66] MARZULLO, K., AND OWICKI, S. Maintaining the Time in a Distributed System. *ACM Operating System Review 19*, 3 (1983), 44–54.

[67] MILLS, D. Internet time synchronization: The network time protocol. *IEEE Transactions on Communications 39*, 10 (October 1991), 1482–1493.

[68] MILLS, D. RFC-1305: Network Time Protocol (version 3): Specification, implementation and analysis, Mar 1992.

[69] MILLS, D. Improved Algorithms for Synchronizing Computer Network Clocks. *IEEE Transactions on Networks* (Jun 1995), 645–254.

[70] MINER, P., MALEKPOUR, M., AND TORRES, W. A conceptual design for a Reliable Optical Bus (ROBUS). *Proceedings of the 21st Digitial Avionics Systems Conference 2* (2002), 986–996.

[71] MORES, R., HAY, G., BELSCHNER, R., BERWANGER, J., EBNER, C., FLUHRER, S., FUCHS, E., HEDENETZ, B., KUFFNER, W., KRÜGER, A., LOHRMANN, P., MILLINGER, D., PELLER, M., RUH, J., SCHEDL, A., AND SPRACHMANN, M. FlexRay - The Communication System for Advanced Automotive Control Systems. *Society of Automotive Engineers (SAE) 2001 World Congress* (March 2001).

[72] MULLENDER, S. *Distributed Systems*. ACM Press and Addison-Wesley, New York, 1994.

[73] OCHSENREITER, W. *Fehlertolerante Uhrensynchronisation in verteilten Realzeitsystemen*. Dissertation, Vienna University of Technology, Faculty of Technical and Natural Sciences, 1997. (in German).

[74] OLSON, A., AND SHIN, K. Fault-tolerant clock synchronization in large multicomputer systems. *IEEE Transactions on Parallel and Distributed Systems,* (1994), 912–923.

[75] OLSON, A., AND SHIN, K. Probabilistic Clock Synchronization in Large Distributed Systems. *IEEE Transactions on Computers,* (1994), 1106–1112.

[76] OWEN, R., AND LOPES, L. Experimental analysis of the use of angle of arrival at an adaptive antenna array for location estimation. In *IEEE Internatnonal Symposium on Personal, Indoor and Mobile Radio Communications – PIMRC* (1998), pp. 607–611.

[77] PARZEN, B. *Design of Crystal and Other Harmonic Oscillators*. J. Wiley & Sons, 1983.

[78] PFUEGL, M., AND BLOUGH, D. Evaluation of a new algorithm for fault-tolerant clock synchronization. In *International Symposium on Fault Tolerant Systems* (1991), pp. 38–43.

[79] PFUEGL, M., AND BLOUGH, D. A new and improved algorithm for fault tolerant clock synchronization. *Journal of Parallel and Distributed Computing* (1995), 1–14.

[80] PUNZ, G. *Distanzschutz nach dem Wanderwellenprinzip für gelöschte Netze*. Dissertation, Vienna University of Technology, Faculty of Electrical Engineering, 1995. (in German).

[81] RAMANATHAN, P., KANDLUR, D., AND SHIN, K. Hardware-Assisted Software Clock synchronization for Homogeneous Distributed Systems. *IEEE Transactions on Computers 39*, 4 (April 1990), 514–524.

[82] RAMANATHAN, P., SHIN, K., AND BUTLER, R. Fault-Tolerant Clock Synchronization in Distributed Systems. *IEEE Computer 23*, 10 (Oct. 1990), 33–92.

[83] RAPPAPORT, T., REED, J., AND WOERNER, B. Position Location Using Wireless Communications on Highways of the Future. *IEEE Communications Magazine* (Oct. 1996), 33–41.

[84] REED, J., KRIZMAN, J., WOERNER, B., AND RAPPAPORT, T. An Overview of the Challenges and Progress in Meeting the E911 Requirement for Location Service. *IEEE Communications Magazine* (April 1998), 34–37.

[85] SCHMID, U. Interval-based Clock Synchronization. In *Seminar-Report of Dagstuhl-Seminar on Time Services* (Schloß Dagstuhl, Germany, Mar. 1996), p. 7.

[86] SCHMID, U., Ed. *Special Issue on the Challenge of Global Time in Large-Scale Distributed Real-Time Systems* (1997), J. Real-Time Systems 12(1–3).

[87] SCHMID, U. Orthogonal Accuracy Clock Synchronization. *Chicago Journal of Theoretical Computer Science* (2000), 3–77.

[88] SCHMID, U. Interval-based Clock Synchronization with Optimal Precision. *Information and Computation 186*, 1 (2003), 36–77.

[89] SCHMID, U., KLASEK, J., MANDL, T., NACHTNEBEL, H., CADEK, G., AND KERÖ, N. A Network Time Interface M-Module for Distributing GPS-time over LANs. *Journal of Real-Time Systems 18*, 1 (Jan. 2000), 24–57.

[90] SCHMID, U., AND NACHTNEBEL, H. Experimental Evaluation of High-Accuracy Time Distribution in a COTS-based Ethernet LAN. In *Proceedings 24th IFAC/IFIP Workshop on Real-Time Programming (WRTP'99)* (Schloß Dagstuhl, Germany, May/June 1999), pp. 59–68.

[91] SCHMID, U., AND SCHOSSMAIER, K. Interval-based Clock Synchronization. *Journal of Real-Time Systems 12*, 2 (Mar. 1997), 173–228.

[92] SCHMUCK, F., AND CHRISTIAN, F. Continuous Amortization need not affect the Precision of a Clock Synchronization Algorithm. In *Proceedings of the 9th Annual ACM Symposium on Principles on Mistributed Computing (PODC)* (Quebec City, Canada, August 1990), pp. 133–144.

[93] SCHNEIDER, F. A Paradigm for Reliable Clock Synchronization. In *Proceedings Advanced Seminar of Local Area Networks* (Bandol, France, Apr 1986), pp. 85–104.

[94] SCHNEIDER, F. Understanding Protocols for Byzantine Clock Synchronization. Technical Report 87-859, Cornell University, Department of Computer Science, Aug. 1987.

[95] SCHOSSMAIER, K. An Interval-based Framework for Clock Rate Synchronization Algorithms. In *Proceedings 16th ACM Symposium on Principles of Distributed Computing* (St. Barbara, USA, Aug. 1997), pp. 169–178.

[96] SCHOSSMAIER, K. *Interval-based Clock State and Rate Synchronization*. Dissertation, Vienna University of Technology, Faculty of Technical and Natural Sciences, 1998.

[97] SCHOSSMAIER, K., SCHMID, U., HORAUER, M., AND LOY, D. Specification and Implementation of the Universal Time Coordinated Synchronization Unit (UTCSU). *Journal of Real-Time Systems 12*, 3 (May 1997), 295–327.

[98] SEIFERT, R. *The Switch Book: The Complete Guide to LAN Switching Technology*. John Wiley & Sons, Inc., 2000.

[99] SHIN, K., AND RAMANATHAN, P. Clock synchronization of a large multiprocessor system in the presence of malicious faults. *IEEE Transactions on Computers C-36* (1987), 2–12.

[100] SIMONS, B., LUNDELIUS-WELCH, J., AND LYNCH, N. An Overview of Clock Synchronization. In *Fault-Tolerant Distributed Computing* (1990), B. Simons and A. Spector, Eds., Springer Verlag, pp. 84–96. (Lecture Notes on Computer Science 448).

[101] SRIKANTH, T., AND TOUEG, S. Optimal Clock Synchronization. *Journal of the ACM 34*, 3 (Jul. 1987), 626–645.

[102] STEIN, S. Frequency and time - their measurement and characterization. In *Precision Frequency Control* (1985), pp. 191–416.

[103] SURI, N., HUGUE, M., AND WALTER, C. Synchronization issues in real-time systems. *Proceedings of the IEEE 82*, 1 (January 1994), 41–54.

[104] THAMBIDURAI, P., FINN, A., KIECKHAFER, R., AND WALTER, C. Clock Synchronization in MAFT. In *Nineteenth International Symposium on Fault-Tolerant Computing - FTCS-19* (1989), pp. 142–149.

[105] TROXEL, G. *Time Surveying: Clock Synchronization over Packet Networks*. PhD thesis, Department of Electrical Engineering and Computer Science, Massachusetts Institut of Technology, May 1994.

[106] VASANTHAVADA, N., AND MARINOS, P. Synchronization of Fault-Tolerant Clocks in the Presence of Malicious Failures. *IEEE Transactions on Computers 37* (1988), 440–448.

[107] VERÍSSIMO, P., AND RODRIGUES, L. A posteriori Agreement for Fault-Tolerant Clock Synchronization on Broadcast Networks. In *Proceedings 22nd International Symosium on Fault-Tolerant Computing* (Boston, Massachusetts, Jul. 1992).

[108] VERÍSSIMO, P., RODRIGUES, L., AND CASIMIRO, A. CesiumSpray: a Precise and Accurate Global Clock Service for Large-scale Systems. *Journal of Real-Time Systems 12*, 3 (1997), 243–294.

[109] VIG, J. Acceleration, vibration and shock effects - ieee standards project p1193. In *Proc. 1992 IEEE Frequency Control Symposium* (1992).

[110] VIG, J. Quartz crystal resonators and oscillators for frequency control and timing applications - a tutorial. U.S. Army Communications-Electronics Command, Attn: AMSEL-RD-C2-PT, Fort Monmouth, NJ 07703, USA, September 1999.

[111] WALLS, F. Environmental sensitivities of quartz crystal oscillators. In *Proc. 22nd Ann. Precise Time and Time Interval (PTTI) Applications and Planning Meeting* (1990), pp. 465–477.

[112] WEIGANDT, T. *Low-Phase-Noise, Low-Timing-Jitter Design Techniques for Delay Cell Based VCOs and Frequency Synthesizers.* PhD thesis, University of California at Berkeley, 1998.

[113] YAMASHITA, T., AND ONO, S. A statistical method for time synchronization of computer clocks with precisely frequency-synchronized oscillators. In *Proceedings. 18th International Conference on Distributed Computing Systems* (1998), pp. 32–39.

[114] YANG, Z., AND MARSLAND, T. Annotated Bibliography on Global States and Times in Distributed Systems. *ACM SIGOPS Operating Systems Review* (Jun. 1993), 55–72.

[115] ZAGAMI, J., PARL, S., BUSSGANG, J., AND MELILLO, K. Providing Universal Location Services Using a Wireless E911 Location Network. *IEEE Communications Magazine* (April 1998), 66–71.

Appendix

The following table gives an incomplete list of available standard devices for different Ethernet technologies.

Fast Ethernet devices			
Vendor	**Device**	**Type**	**add. Interfaces**
AMD	Am79C972	MAC	PCI, MII, GPSI, Eeprom IF
	Am79C975	MAC + PHY	PCI, Eeprom IF, 10Base-T, 100Base-Tx/Fx
	Am79C976	MAC	PCI, MII, Eeprom+Memory IF
	Am79C978	MAC + PHY	PCI, MII, Eeprom IF, 10Base-T
Hitachi	SH7615	DSP + MAC + ...	DSP, 32-Bit Bus IF, MII, several GPIOs
Intel	21143	MAC	PCI, MII, Eeprom IF
IBM	PowerPC 405GP	CPU + MAC + ...	CPU, PCI, MII, Memory IF, serial ports, 32-Bit Bus IF, several GPIOs
Motorola	MPC8265	CPU + MAC + ...	CPU, PCI, MII, several GPIOs, UTOPIA, serial ports
National Semi.	DP83815	MAC + PHY	PCI, MII, 10Base-T, 100Base-Tx, Eeprom IF
NEC	uPD98502	CPU + MAC + ...	CPU, PCI, MII, USB, UTOPIA, serial ports, Memory IF
NetSilicon	Net+ARM	CPU + MAC + ...	CPU, MII, serial ports, several GPIOs
Realtek	RTL8130	MAC	PCI, MII, Eeprom IF
	RTL8139	MAC	PCI, MII, Eeprom IF
SmSC	LAN91C100	MAC	32-Bit Bus IF, MII, Eeprom+Memory IF
	LAN91C111	MAC + PHY	32-Bit Bus IF, MII, Eeprom IF, 10Base-T, 100Base-Tx
AMD	Am79C901	PHY	MII, GPSI, 10Base-T
	Am79C874	PHY	GPSI, 10Base-T, 100Base-Tx/Fx
	Am79C875	PHY	MII, 10Base-T, 100Base-Tx/Fx
Davicom	DM9101	PHY	MII, 10Base-T, 100Base-Tx
	DM9131	PHY	MII, 10Base-T, 100Base-Tx/Fx
	DM9161	PHY	MII, 10Base-T, 100Base-Tx
	DM9162	PHY	MII, 10Base-T, 100Base-Tx/Fx

continued on next page

continued from previous page

Vendor	Device	Type	add. Interfaces
Intel	LXT970A	PHY	MII, 10Base-T, 100Base-Tx/Fx
	LXT971A	PHY	MII, 10Base-T, 100Base-Tx/Fx
	LXT972A	PHY	MII, 10Base-T, 100Base-Tx
LSI Logic	L80223	PHY	MII, 10Base-T, 100Base-Tx/Fx
	L80225	PHY	MII, 10Base-T, 100Base-Tx
	L80227	PHY	MII, 10Base-T, 100Base-Tx
National Semi.	DP83843	PHY	MII, 10Base-T, 100Base-Tx/Fx
	DO83846A	PHY	MII, 10Base-T, 100Base-Tx
Realtek	RTL8201	PHY	MII, 10Base-T, 100Base-Tx
SmSC	LAN83C180	PHY	MII, 10Base-T, 100Base-Tx
	LAN83C183	PHY	MII, 10Base-T, 100Base-Tx/Fx, 100Base-T4
Gigabit Ethernet devices			
Davicom	DM9701	PHY	GMII, 10/100/1000Base-T
LSI Logic	L80600	PHY	GMII, 10/100/1000Base-T
Intel	LXT1000	PHY	GMII, 10/100/1000Base-T
National Semi.	DP83820	MAC	PCI, GMII, Eeprom IF
	DP83861	PHY	GMII, 10/100/1000Base-T

Table 1: Fast Ethernet and Gigabit devices

Next to the Gigabit Ethernet solutions some 10 Gigabit devices with an XGMII interface are already available. For more detailed information contact the WEB Sites from for e.g. LSI Logic http://www.lsilogic.com. A new evolving network processor architecture is the C-5 processor from Motorola, that supports Gigabit and 10 Gigabit interfaces as well. Detailed information is readily available from http://www.sps-mot.com.

Glossary and Abbreviations

The following glossary contains only symbols used throughout the text, whereas infrequently used symbols are explained directly in the context of the description. The same procedure applies for the list of abbreviations.

Symbol	Meaning
α	accuracy
δ_{sw}	transit delay through a switch
ε	transmission delay uncertainty
ε_d	transmission delay uncertainty due to a physical layer device
ε_c	transmission delay uncertainty due to a transmission cable segment
ε_s	transmission delay uncertainty due a synchronizer stage
ε_{sw}	transmission delay uncertainty due to a switch
ε_{nic}	transmission delay uncertainty due to the network interfaces (data source+sink)
f_s	sampling frequency
G	clock granularity
G_s	clock setting granularity
P	max. synchronization period
π	precision
ρ	drift
σ	oscillator stability
u	rate adjustment uncertainty

10 Base-T IEEE 802.3 Physical Layer specification for a 10 Mb/s CSMA/CD local area network over two pairs of twisted-pair telephone wire.

100 Base-Tx IEEE 802.3 Physical Layer specification for a 100 Mb/s CSMA/CD local area network over two pairs of Category 5 unshielded twisted-pair (UTP) or shielded twisted-pair (STP) wire.

1000 Base-Tx IEEE 802.3 Physical Layer specification for a 1000 Mb/s CSMA/CD LAN using four pairs of Category 5 balanced copper cabling.

100 Base-T4 IEEE 802.3 Physical Layer specification for a 100 Mb/s CSMA/CD local area network over four pairs of Category 3, 4, and 5 unshielded twisted-pair (UTP) wire.

100 Base-T2 IEEE 802.3 specification for a 100 Mb/s CSMA/CD local area network over two pairs of Category 3 or better balanced cabling. (simultaneous bi-directional transmission)

CSU Clock Synchronization Unit (a clock Asic)

GPS Global Positioning System

NTI Network Time Interface (M-module adapter board)

UTCSU Universal Time Coordinated Synchronization Unit (a clock Asic)

UTC Universal Time Coordinated (official reference time standard)

Die VDM Verlagsservicegesellschaft sucht für wissenschaftliche Verlage abgeschlossene und herausragende

Dissertationen, Habilitationen, Diplomarbeiten, Master Theses, Magisterarbeiten usw.

für die kostenlose Publikation als Fachbuch.

Sie verfügen über eine Arbeit, die hohen inhaltlichen und formalen Ansprüchen genügt, und haben Interesse an einer honorarvergüteten Publikation?

Dann senden Sie bitte erste Informationen über sich und Ihre Arbeit per Email an *info@vdm-vsg.de*.

Sie erhalten kurzfristig unser Feedback!

VDM Verlagsservicegesellschaft mbH
Dudweiler Landstr. 99   Telefon +49 681 3720 174
D - 66123 Saarbrücken   Fax   +49 681 3720 1749
www.vdm-vsg.de

Die VDM Verlagsservicegesellschaft mbH vertritt

Printed by Books on Demand GmbH, Norderstedt / Germany